D1168878

What I Think I Did

Also by Larry Woiwode

What I'm Going to Do, I Think

Poetry North: Five Poets of North Dakota

Beyond the Bedroom Wall

Even Tide

Poppa John

Born Brothers

The Neumiller Stories

Indian Affairs

Acts

Silent Passengers

Aristocrat of the West

What I Think I Did

A Season of Survival in Two Acts

Larry Woiwode

A Member of the Perseus Books Group

"Intermission" appeared in substantially different form as "A Fifty-Year Walk," in *Things in Heaven and Earth,* published by Paraclete Press, PO Box 1568, Orleans, MA, ed. by Harold Fickett, 1998; in edited form in *Books and Culture,* Nov/Dec, 1998; and in *The Best Spiritual Writing,* 1999, ed. by Philip Zaleski, HarperSanFrancisco, 1999.
"Jawbreaker Layers" appeared on the Internet literary journal *Archipelago,* Vol. IV, 1, at www.archipelago.org.

Copyright © 2000 by Larry Woiwode

Published by Basic Books,
A Member of the Perseus Books Group

All rights reserved. Printed in the United States of America. No part of this book may be reproduced in any manner whatsoever without written permission except in the case of brief quotations embodied in critical articles and reviews. For information, address Basic *Civitas* Books, 10 East 53rd Street, New York, NY 10022-5299.

Designed by Victoria Kuskowski

FIRST EDITION

Library of Congress Cataloging-in-Publication Data
Woiwode, Larry.
What I think I did : a season of survival in two
acts / Larry Woiwode.
p. cm.
ISBN 0-465-07848-6
I. Title.
PS3573.O4 W47 2000
813'.54—dc21 00-020993

00 01 02 03 / 10 9 8 7 6 5 4 3 2 1

Foreword

Memory is a magpie after chips of colored glass and ribbon rather than the upright accuracy of objective sequence. So to restrain my memory's tendency to seek the tawdry tidbit and improvise from there, I have referred to correspondence, notes, and notebooks from the time—as in the long talk with William Maxwell about publication. This was recorded verbatim right after (as far as memory allows) and is altered only to accommodate the tenor of the day and the demands of the page, or, as Pete Hamill has put it, "speech on a printed page depend[s] upon rhythmic approximation, and not the exactitude of a transcript."

No statement of substance, however, is added or altered. This is true, too, of the evening with Robert Lowell and other incidents now passing into history. The italicized excerpts from notebooks are exact; I've resisted the temptation to improve on them, for what they reveal also of a tenor of a time and mind. If I did not speak to or offer the book to anyone living who plays a substantial role in it, that person receives a pseudonym, in the assumption that they prefer anonymity. This is not so for public figures, whose lives are hardly their own. Abundant material is included on several, such as Robert De Niro, but I have intended to keep confidences.

I feel I must publicly thank the two who brought the book about: Don Fehr, a stranger to me when he called out of the blue, on the exact day when I was worrying over what to do next, and proposed a memoir, setting me on a course that has provided me with more joy, perhaps, than any book so far; and my wife, Carole, a private person.

She mistrusted the arena of open revelation and asked to be set in the background, yet encouraged my daily work. I've complied with her request as much as possible while telling the story I must. I thank her for her encouragement and the restraints imposed by her request, which helped me enter the heart of this story; and I want to tender the acknowledgment that although she may appear as a shadowy figure called "my wife," as she has asked, she remains for me, as she has for three decades, at the forefront of the unfolding contours of our life.

A further thanks, too, is due her and our children at home, for the jobs, tasks, chores performed while I was pasted to a computer screen (or, worse, was physically present but unreachable), including the meals that appeared as magically as in "Wings of Words," now from daughters' hands; and for the errors of reference and fact—memory again!—that all of them caught, each in his or her own way, enabling me to bring this nearer my original conception: a reflective rearrangement of actual events.

For Ben & Newlyn

and our grandchildren

Medora, Camilla, Christian

Afterwards, lying in the sleeping bag, I tried to analyze the possibilities. By then I had been through five days of it . . . And as I lay there thinking, I finally asked myself: What are your assets? What might be done that has not already been done?

—ADMIRAL RICHARD E. BYRD, *Alone*

Contents

What I Think I Did

I
Snow with Tints of Then

Jawbreaker Layers

It wasn't my head or even my hat I forgot but my gloves. This habit of getting words in lines on scraps of paper with pens or pencils or at a keyboard (where I now tap) has my hands so set to the task they turn transparent. So gloves don't register on the drive to Elgin—four miles on gravel, fifteen on blacktop—until I'm almost to town and glance at the seat.

A checkbook where my gloves should be, surfaces edged with a crystalline light that shimmers everywhere from the snow-layered landscape like fire on foil. Just as when you drive up a glaciated mountain above clouds, struggling with the effort to see, so, too, here, you struggle and blink. There is no pollution and the sky is so purged of clouds on winter days that a silver-blue line grips the white horizon, welding the light in place: North Dakota.

In its brilliance a car is a greenhouse.

My mind is off on a race for the right arrangement to a set of paragraphs for a commissioned piece I have in my head. After thirty years of this I'm at the stage where I have to run to the bank once a week to hurry in a sum or shift another small one to keep an account above zero—a nuisance, not a deterrent, and the pressure of that sends an afterthought rolling in like a footnote: *writer's hands extension of mind, so good at unpacking the purl prose forms they're feelers for words, burning to invisibility, sun on snow.*

In any memoir, like the one I'm working on, I want to sidestep self-consciousness and get at *me*. And as often happens when I have work on the planks, I hear, at the border of sleep, the first sentence for it. The rest will follow like a curtain of snow threading its way over plowed fields. Or that's the optimistic trend my thought tends to take and sometimes passages do fall in place in a cascade of inner recognition but more often the work is like shoving a plow single-handed through three-foot drifts.

In the German-Russian village where I grew up my name was normal. Woiwode. I liked its look and sound, those vowels and wings and its trick of pronunciation—Y-woodie. The village was Sykeston, North Dakota, after Richard Sykes, from Cheshire, England. "In 1883 he established the town of Sykeston, erecting a store building and large elevator," a history of 1900 states. "Mr. Sykes retains large land interests in Foster, Wells, Stutsman, and LaMoure counties . . . He has done much toward the settlement of those counties and has much land still to sell at three to ten dollars per acre. Mr. Sykes has made, at a cost of four thousand dollars, a beautiful lake within the town site of Sykeston, which is named Hiawatha Lake, and is eighteen feet deep in places and two miles long and about a fourth of a mile wide. The lake will be stocked with fish and boats will be supplied and the place become a summer resort."

It never did, though people fish on it from boats and the temperature can be 100 in the shade on the 4th of July. When I was growing up the influence of Sykes was evident in an attentiveness to the English language, an influence that also came out of the British culture at Winnipeg. But by the forties Sykeston was largely German immigrants, some of them first-generation German-Russians—Germans the Czarist government persuaded to settle in southern Russia and then (mostly through the Bolshevik revolution) misused or purged. They spoke a Yiddishy low German, interlaced with Russian, so that "no," for instance, came out not *nein* or *nyet* but *net*.

So Woiwode went well with them but elsewhere caused a clamor. I got a glimpse of the generational effect of this when our daughter Newlyn was three and people would ask her name. "Woiwode," she would say, pronouncing it Y-woodie, as the family has for generations, and then she would spell out each letter, as she heard us do, as if each were essential, too, to its pronunciation.

An impediment to the simplest meeting.

Wood, the name I use for restaurant reservations, would be simpler, and for a while I considered a change to that. But for the sake of my father and grandfather and great-grandfathers and the vowels and distinction I first saw in it, I suspect, I did not do that. The name is Slavic, maybe Romanian, according to a scholar who worked on Romania's national dictionary.

"Dracula is the Voivoda of Valachia!" he said, happy to hit me with that. I had an inkling the name was Slavic, to bequeath such Tartar cheekbones, but the ancestors I heard about were German, or German-speaking.

Then the one who set me straight showed up. It was the spring of my junior year at the University of Illinois in Urbana, and I was studying the metaphysical poets, trying to place Marvell in the viewfinder of his mower poems, as in

> *When Juliana comes, and she,*
> *What I do to the grass, does to my thoughts and me, —*

scything apart the poet's state as a scythe scythes down grass. As I meditated on that I almost bumped into a tall woman beside a twiggy hedge, her arms in an X over the books at her chest, smiling at me in amusement. She invited me to the noisy basement cafeteria of the campus YMCA, called the K-Room, for a cup of coffee—I was headed there anyway—and once we sat in a beige and orange booth, with mugs on the table between us, she said, "More embarrassing than asking you for coffee, Larry, is how it shames me to say this: Will you screw me?"

Her fingers were white around her mug but she smiled, baring lipstick-flecked teeth in her mannish lower jaw—the fiery German actress I had seen in scenery-eating roles in campus productions. "I asked my divine director of late who to ask, after he said no—he's too gay to get involved, he said—and he said, 'Woiwode.' So sorry about your awful name, love."

My lungs went flat, *I* did, at her accent that had a tinge of Brit onstage but was the accent of the women of Sykeston. My response is blanked. Later she came to the room I rented at the front of a private house, toting a shopping bag, and sat on my only furniture besides a bed and desk—a Victorian couch with a coronet of maple trim across its crown. She was an East German refugee who made it to England, then found an Illinois businessman willing to pay her passage to America and, with his wife, adopt her, she said. Then he started having her at his office when she was fifteen.

She pulled a banana from her shopping bag and peeled it halfway down, in strips, her blond hair drawn from her domed forehead in a pompadour, a Nordic Doris Day of basketball height. She had an entire bunch of bananas in the bag, it turned out, and after the first she removed her coat; the next, her jacket; blouse, though this didn't seem planned, judging from her nervousness and the grapes—also in the bag—which she ripped apart and held to me in a trembling hand.

She leaned back on the couch, near the end of the bed where I sat, its heavy upholstery raying around her hips in emerald wrinkles, and then stood in heels and paced past my knees as if she were onstage, one hand over a hip as she turned away, then pivoted back, her face streaked white with the emotion that was her forte and cried, "Larry!"

I put a finger to my lips to remind her of the quiet I must maintain, since the elderly folk I rented from sat in rockers on the other side of my wall, adopting a patient rock, it seemed, for my first university infraction.

Her face was so close I saw a glaze like greasepaint over the subtle pimples on her forehead, and she whispered in the accent of Sykeston, "Larry, poor fellow, Vyvoodie is Polish! It is no name a Cherman would want—a lousy name! It was a Voivoda that led the bastard Huns to Chermany, their chief! You do not know the ugly affront it is to a Cherman woman who hears Vyvoodie—they are now frigging little petty officials over there!—or anybody who knows their history! How truly evil were these Vojvodas! Evil, evil Huns, not Chermans! Read Sienkiewicz, love."

At anything untoward I went silent, listening as if my life depended on it, and for me it was an affront merely to hear her say "love." Then she turned grave and teary and told me about the businessman in detail.

I was twenty and mention this not to mesmerize anyone into the goggle-eyed mindlessness pornographers like to incite, nor to warn away anyone who meets a Woiwode, but as a cautionary tale for my children. Hear! The Woiwodes and their interrelated clans, over the generations I've known them (now three, with a glance in opposite directions into a fourth and fifth at both ends) have been susceptible to sexual mishap, sexual misadventure, sexual excess, sexual sin—however you see it, no matter their age. So children, wherever you are when you read this or remember it, heed me.

And this: it wasn't until two years later that the defining conclusion to her visit came, in New York, when the wisest and most clairvoyant of modern storytellers, William Maxwell, said, "If you sit or stand still in one place long enough, even on a street corner in New York, your story will walk up to you." So will yours.

My story is partly about the "Old Country" of every immigrant family, plus the otherworld of Europe, and how I was spared the seduction into its centuries-old maze when a Woiwode sailed to the New World in 1867. Then a century later I encounter a young woman from the Old Country who has been seduced and worse by the New and so seems set on another round of seduction.

Those layers are at the heart of the story I know.

Woiwodes are susceptible to such, I said, as others are susceptible, not *synonymous with*. They also tend to be thoughtful, if not intelligent, and slow to rouse to anger, though once angry they burn red hot, often at injustice. They have served as teachers, federal officers, finish carpenters and plasterers, farmers; in medicine, in the church, the military, and nature conservancy. Younger generations are in banking, vet medicine, nursing, real estate, federal inspection; they oversee commercial building projects, are managers, accountants, in law enforcement, and guarding you against computer scams.

Linked to them is the Thiel family of my Grandma Woiwode, known for its scholars and priests. Only one of all of them has stuck to writing.

As for the inheritance from my mother's line, the Johnstons, from the Norwegians on *her* mother's side, the Hyerdahl branch, is the adventurer Thor, of the *Kon Tiki* voyage and book of that title, and on the Halvorson—

Well! a great-aunt exclaimed, scandalized to read in a genealogy she received from a relative in Norway that a male ancestor "drowned in a vat of beer."

What was so scandalous to her was how this was written right on the genealogy for anybody to see!

From what I know about a tendency on that side, I doubt his death was accidental. He was drinking up the vat and popped. And next the Scots-Irish-Welsh of the Johnston half who tended, like many from the edges of the U.K., toward garrulousness and religious extremism. One great-uncle could barely walk by the age of forty from spending so much time on his knees in prayer—for most of the rest of his family, as it turns out. So from my mother's side that, and the related weakness of a diabetic strain; and from both branches an excellent or else a shaky sense of handling money, along with the compulsive way a Woiwode—most every one a teetotaler—will empty a glass of water in great quick gulps, as if a beleaguered ancestor died of thirst.

Then this:

> **Vaivode** (vekvÄ"d). **Now Hist.** Forms: ¶. 6–7 **vayuod'e-, 7 vayvod, 7, 9 vayvode, 8 vaywode. ß 6 uai-, 7–8 vaivod (7 vavoyd), 7–9 vaivode, 8 vaiwode.** [ad. Bulg. and Serb. *vojvoda*, Czech. *vojevoda*, Pol. *wojewoda*, Russ. *voevoda*, whence also Roum. *voevoda, -vod*, mod. L. *voivoda*, mod. Gr. boeboda (e)] = Vaivode
>
> **a. 1570** in Hakluyt *Voy.* (1599) I. 401 When we should have deliuered him with the rest of his fellowes vnto the Voiuo-

daes officers. *Ibid.*, Kneze Yoriue your Majesties Voiuoda at Plasco. **1599** *Ibid.* II.i.198 Voyuoda of Bogdania and Valachia.

b. 1614 Selden *Titles Honor* 249 That of Vaiuod or Uoiuod, vsd in other parts of the Eastern Europe, being, I think, a Slauonig or Windish word. **1686** W.Hedges *Diary* [Halk. Soc.] I. 232 I went to visit and present ye Voyvode and Musellim of Diarbekeer. **1833** R.Pinkerton *Russia* 111 Now but an insignificant-looking place, though formerly the residence of a Voivod. **1869** Tozer *Highl.Turkey* I.141 The protectorate . . . passed into the hands of the Hospodars or Voyvodes of Wallachia and Moldavia. **1884** W.Carr *Montenegro* 22 By repeated efforts the voivode maintains with difficulty a position on the coast.

g. 1847 S.Austin *Ranke's Hist. Ref.* III. 31 He encouraged Francis I. to keep alive the agitation in Germany, . . . and to support the Woiwode of Translyvania. **1847** Mrs. A. Kerr tr. *Ranke's Hist. Servia* xvi. 303 Amongst those executed before Belgrade were venerable Senators . . . and aged and renowned Woiwodes. **1868** *Daily Tel.* I Sept., To be a prince of its park, lord of its lake, ruler of its river, and woiwode of its woods.

Or so the Oxford English Dictionary says, while next to it Webster's Third states:

vai vode \'vai,vod\ *or* **voi vode** \'voi-\ n -s [*vaivode* fr. **NL & It** *vaivoda*, fr. obs. **Hung** *vajvoda*, fr. **Serb & Slovene**, *vojvoda*, fr. **OBulg** vojevoda, lit., chieftain, fr. *voinu* warrior, soldier (akin to **Lith** *vyti* to pursue, hunt) + *voditi* to lead; *voivode* fr. **Russ** *voevoda* fr. **OBulg**—more at **VIM**]: a military commander or governor of a town or province in various Slavic countries.

While back of all of this I hear the voice of an officious recorder at Ellis Island saying to his table-side mate near the

"In" door, "It would be the same, I suspect, as the way these people pronounce wegtable soup."

2

The beginnings of memory are eidetic, pictorial, images of the essence of a day. No language attached. Then a long mid-phase of learning. Then age, with new avenues the mind lays down. These accumulate fast. With the growing errands the mind must run, the avenues that plumb the deepest become the most important. Gloves and scarves disappear. Or the weather of the inner world draws one in so far it's hard to keep the outer in sight, keep a toehold, a grip on it.

As it has been this winter, merely keeping abreast. Snow to the eaves of buildings, which is bad enough, but the worst is the wind and the way it magnifies the cold. It wears at my wife, pouring in streaming weight over the north of the house, the wall where our six-foot headboard stands, with a force I feel will bear us off into the night. Then strikes in wallops that jerk shrieks from nails as the bedsprings tremble under us in our suspended sleeplessness.

I remember the morning she and I drove in early to the polling place, a bulky brick cube trimmed with rows of vertical windows that give it the appearance of gaining an extra story in mock surprise—the county courthouse, set on a hill in town. We watched a county employee, a woman, use a snow-blower to clear the walk to its double doors, a spangled fountain of snow arching from the candy-red machine across a blue spruce that reached to the second story—a patriotic vision of sorts (we were on our way to vote, after all: November 5, 1996) but I should have taken it as a warning: that much snow on the ground before December.

Twenty below for a week so when the sun appears and cooks the snow to a dazzle you enter a daze similar to the one from a dream that holds you under too long, the dangerous state that undoes polar explorers.

When I wake in Elgin the teller with my check is saying, "Do you think we're going to get that terrible storm?"

I'm almost offended—so absorbed I didn't listen to the car radio, and now this affable woman I enjoy talking with is asking about a storm, after what we've been through?

In a voice so tremulous and phlegm-rattly it feels I haven't used it all day, "No" comes out like a strangled moo—*Nooo*. "No," I say, normal. "No, I don't think so."

I'm wearing a light jacket and flimsy cap and realize her question may be her way of warning me, since natives can be politely oblique. With her knowledge of the details of the financial state of everybody in the county and her ability to face them every day, she's an artful expert at the mode. Her square shoulders shift in an interrogatory way, as if to ask if she's gone too far, then she pauses in her count of cash and studies me from eyes magnified by her glasses, head on.

Outside, exhaust from cars and pickups plumes both sides of the street, all the vehicles running, doors unlocked, not a person at the wheel of any. I run to the store and buy food, sensing a hurry there, as if others know something I don't, and on my way to the car streamers of snow start plastering every surface facing west, including my face, and I think, *Twenty miles!* And her with no wood.

I decide to see the seasonal worker who helps supply our wood. I'll encourage him to deliver a load in the best way—drive to the abandoned farm where his house-trailer sits and hand him cash. We've cut every dead tree on our farm and moved on, with an absentee owner's permission, to a stand of cottonwoods two miles down our snow-clogged road.

What we call our farm or ranch neighbors refer to as a garden plot—160 acres. It is small in comparison to the spreads that extend for miles over the roll of our all but treeless landscape, with a sky that booms pure blue to the horizon on all sides, so that even a seasoned traveler like Peter Matthiessen, when he got in his genteel elegance from a car in our lane, turned in a slow circle, to take it in.

He was working on what would be *In the Spirit of Crazy Horse*—later pulled from bookstores by a lawsuit brought by the Governor of South Dakota—and I had driven him from the Bismarck airport a hundred miles south and west to our farm, below the Cannonball River Custer followed on his way to the Little Big Horn.

Matthiessen paused and stared at our garden with its pool of water at a far corner from the spring melt (after the only other winter with snow this heavy), and said in the mildly Oxonian accent he and George Plimpton picked up at the boarding school they attended together, "You've got shorebirds there." He pointed to a glare of water, the shallow garden pool. "I do believe that one is a godwit. I'll be. Do you have a pair of binocs?"

The gentle man who helps supply our wood, a three-hundred-pound biker with a full curling blond beard and a blond ponytail, lives north of the next town, New Leipzig—one of those Dakota settlements on the other side of the railroad tracks running parallel to the highway, so small that from your car you can see down three blocks of the main street to fields at its far end.

But before I'm there a wind hits and streaks of snow revolve to the horizontal and double in volume, a whiteout. I touch the brakes, blinded, our aging Lincoln seeping cold. I reach for my gloves, and the bare spot, added to the storm's onslaught, confuses me so much I swing off too soon, onto the wrong road. But I keep going, a trait my wife translates as bullheadedness but that registers in me as dislocation. I usually have my directions right and imagine if I drive farther I'll reach the stability of the right place and the panic pressing me on will stop, an awful circle. My father did this, as he aged, and I hated it.

I can turn west in a mile, I figure. Land here is laid out in mile squares or "sections," and the boundaries of each, by law, should be roads but lately are so seldom traveled it seems stagecoaches left the last ruts.

Memory isn't a pilot but a backseat driver who wants control. Story is the pilot, and we follow its course through the present, hearing memory's nagging knowledge of the weathers and roadblocks of the past. Memory's aim is to *be* there, leap the present, persuade us the past is identical to the future, prophetic, our one seat of reference—those blank spaces we slip from to find we've been suspended in the past. That suspension is memory's power; memory *is* imagination. It holds a lifetime store of every angle and declination of experience and sensation and fact we know, besides its tinting of all of those. What we call a memoir is an attempt to tame memory's takeovers into paths we tiptoe down toward truth.

By now the snow sailing in from the west is forming finger drifts across the road, or so I see when I can see, and once I rumble over a few of these I realize I have to turn back. But no place. Earlier snows brim both ditches, bulked up by plows, all joining the falling snow in swirls, and then through the blue-gray blur I see a stab of light to my right and a mailbox goes by too quick for me to stop. I'm entering a turn I must negotiate with every sense alert, a banked ascent, and in a stunning whiteout I see it's a butte, a steep one, and I'm climbing it.

Easy to turn on a grade, once at its top, I think, using the downhill slope. I hit a pillow drift with a wallop, then another, and the car, far from the top, slows and starts going sidewise, tires in the rubbery warble of a spin, and I hear the rhythms of a Roethke poem—of his driving alone down a long peninsula, the road lined with snow-covered growth, a dry snow ticking his windshield, the road going from blacktop to rubble and ending in a rut where the car stalls, churning in a snowdrift until its headlights go out. Which is where

I am, at a dead stop, though the engine and headlights and heater still work.

Then it comes as before, the blankness of black velour, a midnight sky without a star. A point of light appears. It travels across the void, leaving a trail fine as frayed filament. It joins similar trails, millions of them, but the whole host do not lighten the dark an iota. The trails travel through ages, down through who knows how many millennia, until stars appear, the sun and moon, the earth in its aqua symmetry and froth of clouds, and then liquid splashes over a floor as everything rushes forward to a burning marvel I know is light.

Voices are raised, shouts. An icy grip surrounds my forehead and my mother screams.

Push! comes the shout. "Push! The hardest part is over!" My ears gain airy freedom and in the replaying of this I feel a tug of sympathy for my mother, who bore me and will bear me through time, though not long—nine years and three months. But the flash takes too long to catch. The simple truth is I was born in Carrington, North Dakota, into the dark hour of 6:00 A.M., on October 30, 1941.

My name is Larry, not Lawrence (for those who want to improve on her), an irritating diminutive common at the time, perhaps because of Olivier, a Laurence called Larry by admirers. So I used L when I began to publish, but Larry was congenial to the mid-part of life, as she probably knew, although tough to fix on a fellow in his fifties. So I often want to return to where I began as a writer, L.

An undeniable fact about the assembly present at my birth is this: I'm the only one living, able to slip into the salty broth of blood on her naked heat. As all newborns do at birth, as ours did, my wife's hand across a stained and miniature back where ribs fine as wishbones expanded and fell with the panting breaths of all four.

The least trustworthy of the better attributes of any mind is memory.

3

As I sort the first images from my past that feel authentic I'm at an upstairs window of a house on my Grandma and Grandpa Johnston's farm, not quite four, because my grandparents move before I'm that age. Out the window I see our family car. It is squeezed between the house and a granary, as if wedged there, and perhaps I do the wedging, because my parents are inside, trying to leave. An internal picture of them in the front seat as its motor starts causes me to yell, "*No!* I shanged my mind!"

I'm supposed to stay with my grandparents for a spell of days, as my parents travel, but now feel I can't. Somebody calls or runs to the car and my mother's younger sisters, Yvonne and Elaine, enact for their family audience "I *shanged* my mind!" "*He* shanged his mind, is that what he said." "Yes, *shanged* his mind!"

Like all my mother's family, they are flawless mimics, and every time one or the other sees me I hear "Have you *shanged* your mind?"—a question that plagues me when it rises in their voices as I work through yet another revision of a story I told my wife was finished. But then I found that *The New Yorker* had a phrase to cover this, too—when a piece was not only done and taken but the fixing and editing was done also, then it was "Done and done."

The only time Yvonne didn't mention my change of mind was the last time I saw her, in 1992, when she was gray-haired and drinking "gray panthers"—vodka with grapefruit juice—in her brother's apartment in St. Paul, there from L.A. with her husband to visit. She was carrying the cancer she would die from in a year, though none of us knew it, not even her. But I sensed a reserve in her (and not only for neglecting to say "shange," out of charity, as I took it, because she seemed the one changed), and then I saw in her face, not in her eyes or mouth but the bones of her face, the face of my mother, dead for forty years.

Before I get out of the car to check where I am on the slope, as I would with gloves and warm clothes, I start backing toward what appears to be a place to turn. But when I revolve the wheel to enter it, the car slides. I touch the brakes, it slides worse, and I remember how, in the letup in the weather early in the week, first a thaw came and then a rain. The wind has scoured this area of the slope to ice, I see as I get out, then go into the tottering running in place that's so funny to others the second before you fall on ice. But I don't.

The nose of the car rakes northwest, its rear tires at the edge of the drifted ditch. From the trunk I get a grain scoop I put there for an emergency. But the snow is so hard I have to stomp on the shoulder of the shovel to get it to bite. During the terrible cold, as winds shifted from one quarter to another, my son pointed out how the snow was worn to grains, like polished sandy quartz, and it had the weighty heft of sand—its corrugated waves as solid as sculpture, the sastrugi Byrd encountered in Antarctica.

Meanwhile freshly falling snow, clumpy and damp from the day's warmth, plasters the side of the car and clings in a film to my flimsy jacket, while the wind brings the temperature down a degree a minute. My hands are numb, fingers like wood, the last of the blood in them, as it feels, about to squirt out my nails. I jam the scoop in the snow and jump in the car—still running.

You should have known better. The clunky, useless scoop, no gloves, staying on the wrong road, then starting up this hill. I should have backed straight down, even if I couldn't stop, till I was on a flat; should have been more charitable when my wife called the other week to say she was backing out a drive and slid on ice into a ditch and, once she was home, I should have forgone my lecture on how to manage on ice, all before I went out to look—she'd hooked the steel post of a highway sign on the way down and scraped one whole side of the new finish we had a body shop apply a month before—and then I went back in and said, "We might as well drive the damn thing over a cliff."

"Dad!" our son said. "That's no way to talk!"

I get out and dig and my mind fills with Ruth, in this helpless tumble of our children I experience with each task—Newlyn, Joseph, Ruth, Laurel, that order. Ruth always busy at a task, Newlyn, too, but with her, the oldest, a sense of how her work was a duty to hold the family together, when I wasn't, Ruth so geared up she's a power I can't identify, though she helps with what she likes. As when I was digging holes for a hitching post for Newlyn (nine years older) and Ruth, four, pulled piles of dirt back with her hands, helping, and when I was down three feet with the posthole digger she wanted to be in the hole. So I lowered her until only her eyes and a crown of white-blond hair showed, and she giggled and whooped and had me call her mother to see.

I dig till my ears feel they're going up in flame, then hear a deep-throated sound like a tractor in the distance. Snow is driving from the north now and might have caused a hallucination with its sudden shift that's dimmed the last light. I've lived in Manhattan, Brooklyn Heights, Chicago, St. Paul, a manor on the Hudson, in suburban and rural places from New York State to Michigan and Illinois, and now, in this corner of North Dakota, I stand on what could as well be the last hill at the end of the world. I see nothing but white with white over it and more white pouring in.

My great-grandfather Charles was smuggled out of Upper Silesia in 1867 by his father, John, because he was at the legal age for military conscription in Germany, ten. John wrapped him in a feather bed and carried him over his shoulder onto a ship bound for America, where Charles said he was sat upon and shoved around during the journey, but kept silent.

And, oh, how I imagine him encased in that mattress, the layers of his mind bright and varied as a jawbreaker's, each one wearing away to the proximity of exposure, not knowing which "I" is I, sorting colors of entrapment, invention, projection, while he keeps his cunning silence, an exile now. That silence and sense of exile he passed on.

But managed for his father on this side, filing a homestead claim in Dakota Territory, in 1881, before North and South Dakota were states. That came in 1889. The homestead was in the Red River Valley, three hundred miles from our present farm, which is not a family place, as some think.

My wife and I are children of the sixties, or anyway we were under thirty at its peak—armchair dreamers of an ecological Eden. We would find a place in the country and by our crops and animals and a system for generating electricity would become self-sufficient, returning organic balance to the land, for its sake and the sake of the life on it, especially wildlife, my wife's love.

I never thought I'd return. My wife is from Oregon and we visited the Pacific Northwest first, spent a summer in Nyack, and then tended west: Michigan, Wisconsin, Illinois, a summer in St. Paul. After two years in Chicago we took a tour of the West—Montana, Wyoming, Colorado, Idaho, New Mexico, and found aspects of each state we liked but not the right place. Along the way we stopped to see the only relative I knew in North Dakota, near its western border, my mother's sister Elaine, and Ralph, her husband, and with him attended a rodeo at Sentinel Butte.

Back in Chicago, it was western North Dakota that seemed to retain a sense of the Eldorado Americans pursue: the frontier.

We had joined a Presbyterian denomination and found it had one church in North Dakota, in its southwest corner, in Carson. With a compass I drew a circle with a radius of fifty miles around Carson, the focus of our hope, and we contacted a realtor in the area. Months later, when the realtor called with a farm that seemed a possibility, I was too ill to travel, and my wife went to see the place with her father.

Both her Peterson grandparents are from Norway and met in the U.S., so she is third-generation Norwegian on her father's side. The pastor of the church that was the point of my compass was a Peterson, a coincidence my father-in-law never got over, and perhaps it was that or the lilting talk of residents (sixty percent are Scandinavian) that was the clincher.

According to my wife it occurred for her at the midpoint of the drive from Bismarck in a rental car, before she saw the farm, as she swung around a curve on a hill above Flasher, where the Missouri breaks give way, and saw the countryside below, fields interleaved with pastures, buttes bunched in blue mounds in the distance, and her ancestry sprang to the surface.

She was at the wheel, her father beside her, and they saw what their ancestors—mariners and farmers above fjords where the sea exploded in cataracts of foam—saw when they crossed the ocean and came into this country.

My wife felt it through her body in a way that marked this as the place, she said. She sometimes still says, as we round the curve and look out on the land lying below to the horizon, "That's the West" or "It was right here." If I should happen to say, as I have, "Here's your West," she'll merely go "Mmm," letting me know in her gently sensible way that I have no right to appropriate her vision.

Two things happen at once. The snow parts and I hear and glimpse what appears to be a tractor to the west, judging from the round bale gripped in its bucket—now enveloped—and then a pair of lights trundle to a glow behind. They pause, as if trying to see me through the streaking fuzz of pouring flakes, then swing off, past a mailbox they light up, the one I passed.

I get back in the car, flinging dripping snow, but hopeful: two vehicles, lights. Frostbite has my fingers feeling I've held them over a fire long enough to reach rare. I decide to head for where the lights went and as quick as that the storm lifts. I see the shape of a man inside the cab of what I thought was a tractor but is a Payloader, busy on a run from a stack to a corral with a round bale in its bucket, trying to beat the storm.

Headlights reappear from close behind and it turns out to be a farmer in a pickup who lives off the road, his face the color of cowhide, so finely seamed it looks liked tooled leather. He went back for a towrope, he says, but why in hell am I on this road? Am I from out of state? No. Why, any fool knows this road is never plowed past his place once it gets bad. Where am I from? I tell him and he shakes his head as if it's the worst possible place anybody could imagine.

He hooks a tow rope of thickly braided yellow nylon to the bumper hitch at the back of my car and to his pickup hitch and tells me to get in. I drop it in reverse and give it the gas, as he does, but sit like deadweight while his pickup slides to one edge of the road, then the other, back and forth, as if it has straight left and right down pat, and no more. I watch him back up, giving slack to the rope so he can try a freeing jolt, and hop out and warn him how a friend broke a tow rope trying that on our heavy Lincoln.

"Get in," he says.

I do, he applies his jolt, the rope sails off like a slinky snake with jet assist, and I touch my pocket, glad for the cash. But only a pin holding a hook at the end of the rope has pulled loose. He sits on a running board on the driver's side, in the lee of the wind, to repair the hook. Over his pickup box, across the road where the loader works, I see another vehicle, a bulk gas truck come barreling in our direction from buildings now visible in the thinning snow—all this last-minute activity a blizzard brings on—then suddenly cant to one side and grind to a halt. A man climbs down from its cab and looks around, as if in shame, then trots back toward the buildings.

The fellow down on the running board, whose fingers I've worried about since he pulled off his gloves to grip the ice-crusted hook, says, "It sounds like Ray has his Payloader going. I'll go see."

He's been following the activity by sound, as farmers familiar with an area do. He drives down to the canted bulk truck, pausing as if to assess the loss, then drives toward the buildings and is gone behind heaped snow. All farmyards are like this, a contractor has said; he uses a crane to clean out corrals and feedlots so swamped there's no other way. We have damaged our tractor snowblower on the hard-sculpted snow; you have to ram it to break it up to blow, then the blower piles it in drifts that get others going.

From inside the car I meditate on the way darkness tints the overabundance of white pure blue.

This outpouring of nature, its excess not to be copied, and the energy we expend on it; the time it takes, first, to admit its presence, in whatever form in yourself, and then sift it from a poem or book or, better, your life. In the seventies I put together a book of poems called *Match Heads*, to denote their brief blaze, all trimmed to a few lines. My publisher bought it but didn't want to bring it out until after my next novel—which I presumed was nearly done. It took three more years. By then I had poems of a different sort and saw in them the beginning of a story of the relationship with my wife. I added poems and found that many of the match heads also fit.

A new book came of that, named *Even Tide*, for the way the two in a marriage are evenly tied, or not, and the time of day when healing happened in Jesus' life. My editor and I persuaded each other we didn't like the previous title, *Match Heads*, also a pun, but another problem was manifest. *Even Tide* ran to a hundred poems. I removed a couple of dozen and used brief match heads as prefaces or conclusions to others, and a rearrangement of the order of the poems in the story started to surface, and I remembered William Maxwell saying about a collection of tales, in his whispery voice, "When I was working on the final version, with all the arranging and rearranging it put me through, I reached a stage where I wanted to throw myself out a window."

I wasn't quite there but knew what he meant. I called my editor, Michael di Capua, who was working with a number of poets, and he suggested I get in touch with James Wright.

"So how do I put together a cover letter to a book he probably doesn't have time to read and maybe doesn't want to see?"

"Just call him. Jim loves your book." He meant the "next one," *Beyond the Bedroom Wall*, which was listed so many seasons in the Farrar, Straus and Giroux catalogue it became a source of in-house jokes, along with Harold Brodkey's "first novel," the advance on which had been out for so long without delivery that, as one wag put it, "the interest just on that would pay off the national debt."

I liked Wright's poetry and we had talked at New York soirees—once when he was nominated for an award he didn't receive, and afterward said to me in the booth of a bar in the form of advice, "When something like this happens, what you have to do is make the next one so damn good it blasts the bastards out of the bleachers!" Which is what he did, or anyway he received the prize the next time around, after he quit drinking, as we had heard.

"When I finished my first book I showed it to Wystan Auden," he said over the phone in a soft voice. "He was kind about the poems, I think, but what I remember most was him saying, 'Only 43. You need to cut them down to 43, James. You can't have more than 43 poems in a first book.' I don't know where he got the number, probably from his own first book, but I think he was right and I think I got mine close to that. I'm a little afraid to look."

I'm afraid to look myself, but finally I leave off tapping on these keys and get out of my chair and go to a box hidden behind a recliner I seldom use and dig through books in it until I find a copy of *Even Tide*. I open it, afraid the number will be in the sixties, and I'm somewhat relieved: forty-nine.

Still I feel the heat of shame climb my face.

A blur like headlights grows into spots at my back and swings past the mailbox: the pickup. I get out. A bulky yellow Payloader shows above the snow near the buildings and starts up the lane, then pauses and swivels sideways, as if training its robot eyes on the bulk truck, then swings again and rumbles past it, uphill to me. A driver climbs down a ladder from its cab in a hooded jacket, his face hidden, and hooks a chain to the Payloader scoop, then to my bumper.

"You're a ways from home. You're Woiwode, aren't you."

Not the Usual Obituary

Freed by the Payloader, I'm on my way home, the violet blue thick black, as if Arctic winter is descending. I run in with the groceries and Care's first words are, "Did you get some wood?"

I tell her what happened on the hillside and find my words so mingled with fear that they sound invented. "It suddenly turned bad. Awful! Have Joseph and Ruth called?"

Our son and a daughter are in Lemmon, South Dakota, thirty miles south over the state line. They work a weekly sale at a cattle auction ring, a job our daughter, Ruth, sixteen, first found, because it enabled her to spend the day on a working horse. Over the summer she was apprenticed to a quarterhorse breeder in Lemmon and with the fixed focus that is her forte she has one ambition: to ride and train as many as she can. She's been at the sale ring since summer, and when the manager saw that her brother Joseph was driving her down, once the weather turned bad, he asked if Joseph—in the interim of applying to military academies—wanted to work, too, and he said yes.

Such a sale has no parameters except it lasts as long as the livestock does, and I ask my wife if she knows if it's over, because our children are usually home before dark.

"No, I haven't called. I'm sorry, I should have."

It is over, she finds when she calls, but Joseph and Ruth are helping "pen cattle back"—move them to numbered pens the buyers load from; they'll call back when done.

"Maybe they should leave now," I say, and realize how shaken I am by the hour on the hillside. "It's suddenly terribly bitter out there. Awful!" I try to call John, the woodcutter, and get a woman on the line.

I run to our bedroom to change, the phone rings, and I run out and find Joseph on, calm, with the edge of precision in his voice that steadies me. He says Lemmon is getting an awful snow and now the wind is moving in. "People are saying we shouldn't drive back."

"I think they're right. I sure don't like the looks of this. Maybe you should stay with the Jensens." This is a Presbyterian minister's family, with a son Ruth's age, who offered us their house whenever we want, so Ruth stayed with them over last summer, while working with horses. "Yes," I say, as if answering somebody. "Do that."

"Dad? I'm sorry I didn't get more wood cut."

"Don't worry about that! Stay where you're safe! The Lord be with you."

"Thank you. And with you."

I finish changing, my winter drill: silk underwear I wear this year for the first time like a thicker skin sealing in my body heat; over that, below the belt, quilted Duofold, then flannel-insulated trousers; plain Duofold above, an insulated shirt, a jacket; coveralls covering all; felt-lined boots, two scarves, a red ski mask, a mouton-wool cap, leather wool-lined gloves with inserts of metallic-laced cloth, which your fingers warm, but once they turn cold, so does the metal, freezing them faster.

I feel like a bundled baby barely able to walk and shuffle out to check on what Joseph calls our wood-eating dragon, thinking how the subtle and complex communication between parent and child, brother, sister, husband, wife, in which an entire attitude is communicated in a glance or tone of voice, a sigh, is impossible to convey. Tolstoy himself pauses to *explain* the gesture or the psychology behind it, after introducing it in context.

Equidistant from our house are two pastures. In ours a creek performs the slinky curves of a minor stream and across the road from our mailbox (two hundred yards from the house) is a pasture where circles of tumbled rocks form what people call teepee rings. Both pastures are virgin prairie sod, unbroken, and wild pasque flowers blanket the rises in pinks and violets each spring.

The circles of rock were set around the bottom edges of teepees at winter encampments, to keep out wind and snow, and when the weather is worst I imagine myself in one, with only the thickness of a hide between me and a storm, all available wood and buffalo chips buried under ice-packed snow, with only blankets or bison robes for warmth as the wind picks up to forty miles an hour. And can't imagine how anybody could bear it.

The elderly brother and sister who sold us the farm are children of the original homesteaders, the Berns. For the first few years their family of five—the three children close in age—lived in a one-room uninsulated railroad shack hauled onto the property from the tracks down the road. This is now our living room, and as I sit in it on one of the worst cold nights (shifting this and myself within it) I try to calculate how it could have accommodated a family of five.

And this room, small as it is, was not the Berns' first house. They lived and cooked and slept and entertained and planned and read (all dedicated readers) and held Christmas in this room after their homesteading shack, smaller and made mostly of sod, burned down their first year. A prairie fire.

These terrible conflagrations, usually set off by lightning, could encompass miles and like a tornado level everything in their paths, as this one did, a phenomenon I experienced, on a reduced scale, the year we moved in, when I decided to burn off a stubble field and a sudden wind tore the fire out of control within seconds while I watched the wall of it rise above telephone wires at a fence line. Fortunately I had plowed around the field, so for me it mostly meant getting out of the way on a tractor. But a sudden lateral burn leaped our far fence line into a neighbor's pasture and I had to use the tractor with a disk attached to extinguish that.

I know about the Berns' fire in detail, because their oldest, a daughter, Enid, wrote the history of the county and then, for the State Historical Society, a series of pieces on the history of homesteading and her family's place in particular.

"You're the second writer to live here!" she exclaimed in her buoyant but not-to-be-disputed tone of authority. She had been a one-room-school teacher and then a public school administrator for forty years. She had taught with my father in Sykeston four years—"from the time you were born that fall till you were four," she told me. "What I remember about you is how you liked to color. Every time I came to your house, you were lying on the floor, coloring in a coloring book or on a piece of paper your mother gave you."

When I try to pry up a memory of this, using the budlike row of colors in a Crayola box, all I see are the colored pages of comic strips, or funny papers, as we called them, that my brother Dan and I hold over our laps. A sensation like electricity passes over me when our arms brush, from the hypersensitivity of that age, perhaps, or because we're brothers only a year-and-a-half apart, as if our proximity and touch is tactile proof of the wordless well of closeness below our family. The colored paper rustles and I remember how, during one of these sessions, a marvel fell over us like a prickly rain of electrons. We could read.

Today, in the temporal present where I tap this together from drafts and notes, a roadblock arrives, a triptych of my family—a studio photographer's paperboard foldout, taken two years before we moved from North Dakota. My mother and father bulk up in the center frame, sitting, seen from the waist up, her lips pursed and her body heavy from nursing, with "Mr & Mrs Everet Woiwode" written beneath in Palmer penmanship, one T of my father's name clipped; my brother Dan to the left, with his arm around Mary Lois, the nursing baby, both their eyes wide; my younger brother Charles and I on the right, smiling like cats in a fish store.

This has come from a nun, on an afternoon when I've stopped in my pickup at the mailbox on my way to the bank, pulling book envelopes and letters in through its window.

"My mother is cleaning and hoping to get rid of some things," she writes. Her parents were family friends during our last years in Sykeston, she adds, the godparents of Mary Lois—this I remember. I lay the picture, which I've seen, on the seat beside my gloves and open a newspaper clipping that has fluttered from it. There in newsprint, "Thursday, March 15, 1951" rests above a column titled SYKESTON, and below, in smaller bold "(Last Week's News)" is this:

> Just received the obituary of Mrs. Everett C. Woiwode of Manito, Ill. . . .
>
> On Jan. 19th, Mrs. Woiwode was taken ill and upon arrival at the hospital at Pekin, a nearby town, her unborn baby was pronounced dead and a Cesarean was performed. She gained strength rapidly and by the third day her pulse and temperature were normal and she was feeling very well. On the fourth day the doctor became concerned about her kidneys but they didn't think there was immediate danger. From then on she apparently improved and when her husband visited her the evening of Jan. 25th she seemed very well and talked a good deal, and before Mr. Woiwode left the doctors assured him that she was better

> and there was nothing to fear, her kidneys having improved and temperature and pulse normal. About 10:30 that night he was called back to the hospital, she having become very ill. Besides the three doctors already there, three more were called. They recommended taking her to the Methodist hospital in Peoria, where there is one of the very few mechanical kidneys. The next morning at Peoria, she began responding again and by Jan. 28 her condition was much better and at times she seemed to recognize things going on. She seemed to be holding her own and her pulse remained good until at 2:30 P.M. Jan. 30th she very suddenly passed away, before her mother and an aunt, who were in the hospital, could get up one floor.
>
> Mrs. Woiwode, nee Audrey Johnston, was born . . .

That's all I read. I pull off down the road, angry that this is the first I've heard of this, that I've waited this long to discover it, when at the time, fifty years ago, I had to piece together a story with details I overheard or saw, some of it as slippery as if slick with blood. Before I get to the blacktop a reaction like a cough comes, then tears spatter my jacket and I grieve for her all over again—the misty winter countryside a white swim I suck with seawater down my throat.

Back home, chastened, wordless, I hand the packet to my wife. She reads and turns to me with empathetic eyes, the look that attracted me to her thirty years ago. "I didn't know she rallied so much!"

"It's the first I've heard that."

"It's obviously written by somebody who knew her—not the usual obituary."

I take the packet to the building where I work and sit at a desk like a kitchen counter, not sure how long the stupor lasts before I reach with the hand from a dream for the clipping, unable to backtrack:

> Mrs. Woiwode, nee Audrey Johnston, was born in Moorhead, Minn. Nov. 9, 1916, the daughter of Frederick and Lydia Halvorson Johnston. On Dec. 22, 1938, she was married at Wimbledon to Everett Woiwode, who survives. Also surviving are five children: Daniel, 10; Larry, 9; Charles, 6; Mary Lois, 4; and Marcie, 2; her parents; 2 brothers, Raleigh Johnston of Park Rapids, Minn., Wendell Johnston of Fargo; and two sisters, Mrs. Elaine DeFey of White Earth, N.D., and Mrs. Yvonne Jarvis of Inglewood, California.
>
> Funeral services were conducted by Father Sommerfeld of Sykeston in the Immaculate Conception church of Manito, of which she was a member, and interment for her and the infant, Rita Mary, was made in that cemetery.
>
> Mrs. Woiwode was very well known here, after 11 years of residence in Sykeston, and had a great many friends. She was a jovial, kindly, friendly little lady of many talents which she used for the good of her family, her friends and the community.

This parting of flesh at the center of generations. I remember watching from a barn door rain fall over a freshly plowed field and seeing how much silver is in black. And because silver is so close to white I felt the two could be reversed, the world never again what it was. My black mother comes walking toward me in a white gown that was her funeral dress but now is her new life.

2

What I noticed when we moved here, on Labor Day, 1978, was the silence. Any writer would. The house was empty, our furniture hadn't arrived, and I spent the days of our wait re-painting the downstairs rooms off-white. Each of our children at the time, Newlyn and Joseph, had bedrooms upstairs—ours is on the ground floor—and we all slept on carpets in sleeping bags, feeling the outdoors go from chilly down a further notch each night to cold.

We were having a new furnace installed and the plumber hadn't arrived yet; no refrigerator was moved in, not even that purr, and as I lay awake I felt my hearing extend in widening rings to locate one sound. Most people live in a state of perpetual noise. I did, as I discovered once we moved from New York, then Chicago. I'm sensitive to the way noise scatters sentences. It can break up even more dimensional matters (memory and its floes) as readily as a blacksmith banging an anvil. I lack stillness at my center, as I was learning—a funnel of rest I can slip down and lie at the still point of the turning world, as a poet who should know put it: a meditative center where now and again, at last, I can rest.

At first I was sensitive only to outer noise and when a story was published in 1970 in one of those quarterlies that request a seemly bio, I said that I had been traveling the U.S. with my family, trying to find "a place quiet enough to work with some permanence"—so I'm quoted. By then I had read everything of Nabokov in English and what I remember saying was that, like Pnin, I kept moving for "sonic reasons."

I had hardly received copies of the quarterly when a letter came from John Cheever. He was moved by the story, he said, which was about the disintegration and suicide of a father, as viewed by his son, but he also responded in a fatherly way to my odyssey for quiet.

Once on the farm, whenever I had a question, Enid Bern, the historian, always had the answer, until her last year, when she was ninety-five. Whatever I've asked her surviving brother, Ivan—how deep is the well; how far down is the buried pipe that runs from there to the barn; is the propeller in the rafters of the granary an airplane's?—he's been able to give an answer.

"No, the propeller was an extra for our wind-charger."

This was a 32-volt DC system they sold, with a variety of appliances, to a family from Montana at an auction they held two months before we moved in. I flew out to attend and bought some equipment essential to any farm and ran out of money, and as I followed the crowd a couple walked up, my age, I thought, and the woman, who may have been younger, said, "We're your new neighbors. No, wait! *You're* our new neighbor." A duskily seductive alto. "I'm Valeria."

She shook my hand and I found hers, though slim-boned, as callused as a metal-worker's. "This is Kenny." His grip popped bones; talkative, assertive, with a windy love for his talk. After the auction, as I sat in the kitchen of the house soon to pass to us, drinking lemonade, Ivan and Enid told me that Kenny Kibbel, of Kenny and Valeria, farmed their land for several years and was a person I could trust. Alvin, his nephew, lived to the west; it was his yard light I saw in that direction.

Kenny's brother Bryce and his wife and two children were caught three years ago in a flash flood. Kenny had left a washtub out in the open and the rain left eight-and-a-half inches in it; runoff filled our pasture, covered the road, and started across the lawn toward the house. Bryce and his wife and children were away for the day and when they stopped in Mott somebody said, "Maybe you better stay overnight."

No, they had chores. They drove out Highway 21 to Burt and turned on the gravel road that runs south past the Berns' to their house, a mile-and-half farther. Enid had taken a family picture of them days before, beside separate shots of their daughter, turned this way and that in studio poses—her grade-school graduation picture. The force of the water as it came off buttes a mile long washed a ten-foot culvert from under the road, this a couple of miles south of Burt, and a driver who came by earlier, when it was still light, turned back, thinking, *Somebody should put up a warning sign*. Bryce must have thought it was water covering the road. None of the family could swim. They found the children first, then his wife, and two days later his body, far down the Cannonball, so bloated that Kenny later told me he was set outside the church in a sealed container for the funeral. All four dead.

That was one introduction to this place.

3

In the fall of 96, before the worst winter in collective memory came, an outdoor woodburning furnace seemed a good idea.

We had, in fact, put money down on one. In September I asked my wife if she would call the furnace manufacturer, in northern Minnesota, to see if our deposit still existed, and the person who came on said a load of furnaces was leaving for Oregon in two weeks. There was room on their truck for one more. Did we want them to load ours?

So on a day of streaming rain of the kind we get in September, a semi-truck-flatbed with a pup hooked behind (those shorter trailers that semis tow), the whole trainlike rig loaded with ten furnaces, came blatting up the incline at the head of our lane, pouring out black smoke, and on up into our circular drive and around to the house.

We were somewhat prepared because the driver called from town. I went out in one of those over-the-shoulders slickers that are rubberized rectangles with a hole for your head. The row of furnaces looked like outdoor storage sheds, sheathed in green or brown or black corrugated steel, but each had a smokestack emerging at the rear and rising above each roof—all these black-clad and shining in the rain. They were different sizes and ours in the medium range. "Weighs a ton, empty," the driver said. "Add a couple more when you fill it with water." He extended hydraulic legs from the flatbed to steady it and I saw mud peeled back from his tires like sod from a plow and felt bad for my wife, who cares about the looks of our yard and drive. She stood off, her hair streaming water, curling with the moisture, and kept her camera as much as she could under her poncho—an elderly Pentax that has kept track of us for years.

The truck was equipped with its own crane and once the driver knew where we wanted the furnace, next to the yard light on the other side of the drive, he ran a hook from a cable on the crane into a ring on the furnace roof. "You get ahold of it," he said, "and keep it from swinging about and bumping his mates." A Canadian. As the crane whirred my slicker rattled so loud in the rain I felt it could be my bones under the crushing weight, and then our son, Joseph, stepped out, zipping up a light jacket, and did most of the guiding. We eased it to the exact spot and twisted it true, its door facing south.

The door was black steel and opened on a firebox four feet high and five deep, this surrounded by a water jacket that held 400 gallons. I knew the specs but it was another matter to see it close up, rocklike in the rain. I opened the door—a full-length chrome handle with hooks above and below to hold it airtight—and saw the firebox was packed with cardboard boxes.

"Those are the fittings and pumps and the rest to hook it up," the driver said.

"Where's the underground pipe?" This was a high-tech poly hose from Europe that had cost us, I felt, a mint. The furnace and the packed boxes inside had inspired in me that jangled grasping strand of American consumerism: *greed*.

"Not in the furnace?" he asked.

"It's too full."

"Then here." He opened the steel door of a compartment below the flatbed and handed me a coiled gray hose. I gripped it to my chest like gold, that heavy. "Oh," he said. "I see another. Yours, too." He hung another loop from my arms and I was barely able to grapple with the two.

"Do you need help, Dad?" Joseph asked.

"No!"

"And this," the driver said. He added another and I staggered toward the furnace, leaning backward, and knew I couldn't make it, quite. I let them fall in a tumble at the base of the light pole. I was embarrassed, something amiss under my poncho. In my hurry to meet the truck I stuffed my reading glasses, which I need to work at the computer, into my shirt pocket. I reached in and pulled out a mess. Both lenses were popped from the frames, fixable, but the frames themselves smashed beyond repair. I didn't have a backup set and had to fasten the lenses over the holes of a pair of old frames they didn't fit into with scotch tape. *Greed.*

We planned to heat our house and two outbuildings with the furnace. One building is where I've worked since 95, with a greenhouse attached to it, the other a granary once tipped on its side by a tornado that tore away the garage next to it, leaving a cement slab. The tornado hit the year before we moved in, and our first summer on the farm my wife's parents came to visit and I asked my father-in-law if he thought we could right the granary and get it on the slab. He was a tool-and-die maker who, with his skills and engineering sense, devised and patented several innovations, among them a latch for caskets, and went into business on his own. But he preferred plain work to management.

He found a length of cable and passed it like the string for a gift around the back of the tipped building, close to its roofline, and fastened the cable in place with spikes he hammered in beside it and bent over. Then he drew the two cable ends together at the front of the building and bolted them together with a clamp.

"Get your tractor," he said, and I backed an aging Minneapolis Moline I'd bought at Bern's auction to the point where the cables were clamped. We hooked them to its drawbar with a clevis, and I put the tractor in its lowest gear and eased forward. The granary rose, groaning, then slammed down upright. He arranged pieces of board under the cable strung across its back, to bear the weight of it better, then pried up the two front corners and had me place fenceposts, rollers, beneath them. With only a few pauses to add more posts as rollers, then a shift to the angle of the tractor's pull, we got the building true on the slab.

He stood to one side and surveyed it, fists on his hips, with the fit build of a thirty-year-old, although he was pushing sixty, and I thought, *I could work with him*. Which was the closest I could come at the time to what I meant, since it was my first inclination down this route, which was, *I could get along with him*. Through all this we had hardly spoken, since we understood the basics of what we were doing and were familiar with work, and then he said, "There. Now you have an office."

The granary, a government-issue building, fifteen by twenty, was my workroom till 95. That fall my son and I finished remodeling another tornado-damaged building, a pump house—adding a twenty-foot extension and then an eight-foot greenhouse at its front side.

The granary was where my children used to come for "talks." I would ask one of them out or one would be dispatched by my wife for misbehavior, usually Joseph, although

Ruth ran a close second, and I would have to speak to them as a father meaning to keep his household in order. Then Mister or Miss Misbehaver had to sit at one of the desks—doors laid flat all around the circumference of the interior—and work under my supervision.

We taught all our children—Newlyn only partially—at home. My wife bore the brunt of the daily desk work and I attempted to hold "discussions" at the dinner table. She taught each of them to read and write and add and multiply and subtract, on up into advanced algebra, she taught them history and science and grammar, and she tried to teach each of them a language—she knows Russian and Spanish—but for them that was one too many.

Some days she was so filled with their lessons she couldn't think beyond third-grade level, she said, but was able to look on with pleasure when one or another picked up a book that wasn't required and sat down to read. While our children were growing up we never had a TV, and within a year or two of work with their mother they became self-disciplined enough that they merely had to be set on course.

I had other talks with them, when I invited one or the other out and sat them in a chair across from me and told them what points in their behavior I felt were improving, or complimented them on a trait I saw emerging, or an area of study, where a recent effort was obvious, and then I prayed with them. And usually when I stood, done, I felt a sudden grip of arms around my legs. I didn't hold the talks for that and was always surprised at the impact, perhaps because each had his or her own timing and impulses, and I mine. Then the clasp or hug was around my waist, and soon we were gripping one another as adults. That's how quickly they grew.

The advisor for the furnace company, Dennis Filer, a gentle and obliging backwoodsman I was getting to know over the phone, said, "I can assure you your model will heat that many square feet of separate buildings. We have thousands up and going, from Maine to California, and they function so well, even up here—this is America's cold belt, you know—I was using one before I went to work for the company."

Before, to heat the house and building where I write, we used an odd combination of a propane furnace, two wood-burning stoves, a length of electric baseboard, a kerosene heater, and, when the weather turned worse, a milk-house heater with a circulating fan. An outdoor furnace seemed the solution. We would cut heating costs (and eliminate the danger and soot and ash of kerosene and wood) and simplify our lives—one fixture, one fuel—moving us nearer our sixties dream of self-sufficiency.

Just as important was wood is a renewable fuel. Natural gas or propane or kerosene, petroleum products, can't be replaced, along with coal and peat and plutonium rods. Even electric heat uses one or another of those fuels to drive the turbines that produce electricity—and hydroelectric power employs additions or alterations to lakes and streams.

We've planted thousands of trees since we arrived, and will plant more, I presume, and not out of guilt. This rolling plain scraped clean by glaciation is all but treeless; every planting helps hold soil and adds to the wildlife habitat. Across the landscape trees lie where they fall and half-mile tree rows, planted as shelter belts, go dead. The trees around our farmstead were put in sixty years ago, during the Dust Bowl days of the thirties, and most of them, Chinese elm and stunted ash, are slowly (in some instances rapidly) dying. What do you do with a dead tree? Dump it in a poisonous landfill?

We cremate them for the heat.

"It's time, Dad," Joseph says.

We drive in our green pickup to the rows of cottonwood and I open the barbwire gate—glitters of spiny frost on each barb. An owl falls in ashy flight from a high window in a nearby Quonset, the last of an abandoned farm, and pumps its wings almost to touching, once, twice, enough to glide over a hill in the secret silence of snow.

Joseph fills the chain saw with gas and I say, "Let's cut till the tank runs out and call it a day."

"All right."

"Did you see the owl?"

"No, I was thinking." Which occupies him entirely—this concentration of a nineteen-year-old.

"What?"

"Why it is I don't like winter the way I used to."

"Because of all the wood we have to cut," I say, and notice how snow bulges above the bare and wind-polished cottonwood limbs in ridges as thick as the limbs themselves.

"I mean the *season*."

"That's because of this awful winter."

"Probably."

He starts the saw, his turn for this, our meditative silence shattered, and soon the two of us are working as one while darkness swings down on us at 4:00 P.M. as swiftly as the swoop of the silent owl.

The Wood-Eating Dragon Shows Its Fangs

What I remember most from that time, back when words weren't words but tinted suggestions meant as guides to my parents' state of mind, was the ordered spirit in our house at Sykeston. When I enter that house by a trick of tipping myself into its visual surroundings, I rest on the source of my writing, the anchor that holds me to one area I know for certain as I let whatever I want to say fill a page so I can assess its reality.

If I step out the door of the house I stand between two elements that govern the universe: blue and green. The sky and plain are all I know and it is against that cyclorama and on that stage that I'm reared until I reach what may be the most meaningful phase toward maturity—when I use words as an attempt to please my parents.

Once I begin that, I feel suspended between them, too, in addition to the blue and green, or heaven and earth. Each parent represents one or the other, I believe, and I try to work out in myself their placement. But I run into a hedge of difficulty, like the yellow-blossoming caragana enclosing our yard, as I come to understand that every word of mine has a distinct but somehow separate meaning in the realm each one inhabits and (this is what I have to sort) signifies an allegiance to one or the other.

Each word does.

Later I tried to transfer an edge of that preoccupation to a pair of young characters, brothers, in a novel:

> It's difficult for us to fit a word to something it doesn't sound like, especially for him [the narrator's brother]. If the word doesn't match the part of the thing it's meant for, or the thing itself, it slips off the, ah, whatyoumaycallit, you know, that shiny bending thing, wide at the end but flat, for slipping under and lifting—and we learn to use the names that others do, when we do (spatula for that, for instance, now in Daddy's big hand, turning over eggs crinkled at their edges), out of habit.

I can't vouch for the accuracy of a detail once it has wound its way into fiction and reentered my memory by the route of words. But since both parents were, in real life, let's call it, English teachers, you might imagine the verbal habits each of them drilled into their children.

Meanwhile, on my side, I worked at those complications of allegiance language provoked.

He was the earth, of the earth, I decided, my father was, and then I watched a coffin with her body inside sink into the actual earth, the dredged-up dirt heaped around and then heaped on her, and he became the sky I looked to and prayed to and rested under—now entirely confused by this reversal of the two.

Her death, the calamitous event I've tried in different ways to put into the hands of fictional characters, hoping to leave it with them, sometimes returns. But even when the elements of both my parents were present I sensed another, higher use of words, at the age of four, in a literal memory, so I believe, that I put in a nonfiction book, *Acts*:

> I remember waking (or this is the way the sensation arrives) in church and hearing a priest with a German accent declaim in what seemed to me anger, in reference to a passage I now know is from First Corinthians, "Does that mean, wives, that you must submit to him when he asks you to go to bet wid him? Yes!" With the fervor of his yes! I felt my mother next to me stir in the pew, uneasy, then my father shift on my other side, while I experienced at their center my first faint stirrings of sexual intimation—or whatever rough secret it was they shared in their bedroom.
>
> I was raised a Roman Catholic . . . I heard sections of the gospels and epistles read in every worship service as I grew up [every day, actually, attending parochial school] and one of the teachings of scripture is that faith comes by hearing. I was aware I was hearing words out of a realm removed even from the priest.

My years to the age of eight were folded up in a storage trunk and placed in an attic, by a means I plan to disclose, and it wasn't until I was twentysomething that I was able to acknowledge their effect: the most lasting influence on me as a writer.

Because of the people of North Dakota, the spaces of definition between them, and the teaching I received, first from my parents, who embodied what they taught, and then in a parochial school, their influence is my skeleton. The school enforced what my parents taught and once I was able to put in perspective the nun's strictness and forgive, as I'd been taught, the well of scripture overflowed with rivers of life or, to borrow again from *Acts*, it is the Word "that has the ability to break apart every deathly, cyclical activity we have the power to devise, and to make our lives whole again."

Southwestern North Dakota has milder weather than most of the state. Generally. Some winters little snow or a light dusting arrives and is away in a week. A few winters ago we had a stretch of days in the sixties in January. West River Country, as this corner of the state inside the curve of the Missouri is called, is arid, like a miniature New Mexico, even topographically, with its modest hogbacks and buttes—cone-shaped and mesa-topped—and yucca growing wild in ditches. A cactus of the opuntia family, the low-growing, round-stemmed cholla, a form of prickly pear, is overtaking a stretch of our yard.

A few older residents refer to the area as the "banana belt," as it was once called, a promotional scheme meant to attract gullible immigrants, I suspect. A few years ago a local newspaper printed a cartoon, drawn by an ironic young man who is now a sculptor, picturing a pair of men bundled up in Alaskan parkas, breathing steam in the cold, one of them with icicles hanging from his nose, raising an ax over his head while the other holds a peeled-back banana on a chopping block to receive what the caption calls "The Banana Belt."

Our temperature can be thirty degrees higher than the rest of the state—all this, including the aridity, caused by air currents from the Rockies and Black Hills that meet and revolve in collusion above this corner. In June and July the currents cause tornadoes. The one that took away the garage and yet another tore through our farm in a space of five years. My wife and I once stood at our pasture fence and watched a pair of tornadoes, perfect twins, descend from a lime-green sky into a trio of pointed buttes, our landmark to the north, and then spin away into oblivion.

The summer before he died my father-in-law stayed with us a month, and when he went out for a smoke under a Chinese elm whose leaves were continually being bent backward into light-green cones by the prevailing wind, he would come back in and say, "Doesn't this ever let up?"

At a point when I was working on a novel that rose from the story "Beyond the Bedroom Wall," I needed to know if any of my descriptions coincided with the North Dakota from my trunk in the attic, so I drove back with my wife and Newlyn, then four. I was set back in my seat at the palette of the landscape, as resonant as ektachrome under an unpolluted sky. The inhabitants had a resonance of their own. They used few words and with each word they used they intended to mean exactly what they said. So I was brought full circle to that embrace of blue and green, and then the doubling presence of my parents went shifting through as many levels as I could contain in a further embrace.

Story, in whatever form it takes, is our pilot. We are headed somewhere and it's our story that carries us forward in its wake. If I weren't heading toward eternity (as I see it at times), I wouldn't have a story to tell. And you are headed the way you are because your story is bearing you in its direction, as the lees of this one might propel you enough to induce in you a sense of motion.

When you reduce a story to the mileposts of facts, as in a diary or essay, you tend to say too much or too little, either or both. Or as William Maxwell says, "When you explain away one mystery, you only make room for another." The slender or overloaded nature of your content in the case of an essay is sometimes confirmed by an editor—editors of journalism always more assertive about what you should include than book editors or (rare phenomenon) true editors of fiction, those invisible aides who work to shape what is.

Editors of journalism propose that you supply certain material to fit the wants of "our readers," as they perceive them, or "our publication," as they perceive it, at this juncture in its history of want. Journalism's ontological deadline. Not that the rent or the mortgage or the IRS or having food on the table doesn't impose a deadline all its own—the ones that rise out of actual need the most fruitful to creativity.

It may be helpful here to recall that the magazine where the best of journalism appeared for decades, *The New Yorker,* had a fiction department not separate but co-equal, with one common link, Mr. Shawn, as most of the staff referred to him.

One afternoon I was sitting in a chair in front of an oak desk of military width that Bill Maxwell sat behind, when a gentle tapping started at the door that led to the outer office where Maxwell's secretary, Julia, was working. A deferential elfin fellow with the aspect of a messenger, his bald head like a capped buttress to his breadth, stood holding a sheaf of papers at my side, peering at me with his face at an angle, Maxwell already around the desk to introduce him. Mr. Shawn. I was on my feet and had my hand out and Mr. Shawn glanced at it, as if this were curious, and then, shifting what I saw were galleys, took it in his and stared further into me.

"Ah, yes, Mr. Woiwode," he said, pronouncing it right. "Yes, mmm, a-hmmm," in an inner language, it seemed—wearing one sweater over another over his shirt and tie, all under a tweedy jacket, as if to camouflage the monumental diffidence he conveyed. "A wonderful story." Then he was gone.

I turned to Maxwell, already behind his desk in the way he had of transporting his body (perilously thin, I thought, in my misplaced concern for him) from one spot to another—this magical agility of a man nearing sixty! Then I swept my hands down my tattered trench coat, to indicate why I was uneasy. "If I would have known," I said.

"It's perfectly all right. He said he wanted to meet you when he read your last story."

This was an imaginative reconstruction of the night my mother was taken to the hospital, its eleven pages titled "Beyond the Bedroom Wall." That afternoon I had run to West 43rd to drop off the manuscript of another story, wearing over whatever was beneath a black trench coat from college, stained and frayed, its pocket edges torn.

"If I would have known, I would've—"

I held out my hands and stared at the ragged threads dangling from the coat's cuffs, aware (from an astonished glance down during the subway ride) of how similar threads hung in a kind of hula skirt from its hem.

"Don't let it concern you. All he saw was your eyes."

It was a variation on an earlier day when I stood before Maxwell in a new topcoat. I had popped on the coat and a new fur cap to walk from a PR mailing firm on Lex to West 43rd, after a call from Maxwell to the secretary at the front desk. I was working then on what I hoped would be a novel when a phrase that fell on the page caused me to turn aside to a wholly separate scene, following the lines of an incident I remembered about my grandmother, my mother's mother—a story, I was sure—and I typed it up and mailed it to Maxwell, and now I drop in an elevator from that ("Hey, hey!") into the way the checking department at *The New Yorker* was as unsparing about fiction as fact.

In one story I used the phrase "the Willet subdivision in San Jose" and a checker wrote in the margin (with details altered to assuage the innocent), "I find a San Jose in the area in Illinois where the story is set, with a Bergon subdivision in it, and a Hartley Willet in a neighboring town, Manito, which has a Willet subdivision in it, but no Willet subdivision in San Jose. Does author know this?" *Yes!* An awful urge for accuracy animates fiction, as you see across the spectrum, from Welty to Nabokov to Updike to Mary Gordon, a trait that can send a storyteller into a tailspin for a day to confirm a single fact before moving to the next paragraph, rather than plowing ahead and straightening things out later, as a biographer (a factual writer, I assume) told me was his mode.

The strait of a fiction writer goes, I can't set down a sentence if I don't know it's true, because I'm responsible for this. *I'm* on the line, not a "fact" from the *Times* or yet another biography with a slanted interpretation all its own.

That strait or urge was stimulated by every query from a checker, sending sane writers into the overscrupulousness of borderline madness. Which helps explain why it is that a fiction writer feels a tremor like an earthquake when somebody who isn't familiar with the way he or she works says, "I don't have time to read fiction, because I prefer real facts."

The facts of the living and the dead are meaningless until they're caught in the defining net of language and then sorted and set free into the current of a story.

2

I stand in a bookstore in South Dakota, at the end of a scheduled book-signing, next to the only person left other than a cashier, the manager. After two hours of a trickle I want to extricate myself from the manager-fellow's talkative embarrassment, or at least say I've had from hundreds to a half dozen at book signings, and this was not as bad as the low end, so not to worry, people hardly read anymore, and—

At that moment a nun walks up. She wears a matching skirt and jacket any woman might, trimmed with darker piping, but a piece of identical material pinned to her gray hair is the giveaway. For me nuns are messengers. I stand alert.

"Sister!" I say, as I was taught in parochial school, seeing that what she grips to her chest are my heavy books. They were bought years ago, from the looks of their dust jackets, and at the sight of them the manager walks off.

"So you're Larry Woiwode, the author."

"Well." Those are my books, I want to say, uneasy with "author," after years of being a writer, day by day. Now and then I even say that to someone, but wouldn't think of correcting her. Her pink and dandily lined face has the pugnacious aspect of a survivor.

"I recognize you from your pictures. Otherwise I wouldn't know you from Adam."

I sense a jolt of grace or whatever spiritual emptying it is that passes between us (as a husband and wife feel at a surprise sight of each other) and wait for her message.

"I'm Sister Bernadette. I was a Joyce girl, back in Sykeston. I used to baby-sit for your parents."

"Oh?"

"I baby-sat you."

"Goodness!"

My wife, who has been examining books with the downcast countenance of overload from a bookstore—too many books as mere products—suddenly turns and walks over.

"It would have been when you were two or three because our family moved to South Dakota after that."

"Yes. We moved into the house where you had lived."

"Don't get me going on that. You wrote a story about the house in this book"—she joggles the books in her arms—"and got it all wrong. What I've thought about all these years is how I wheeled you around in a stroller. I took care of you and your brother—Danny, isn't it?—when your parents wanted to go off on their own."

"Dan, yes. How was I to take care of?" This wonderment develops when your memory is failing.

"Oh, you know—"

She glances at my wife, so I introduce them. "My wife," I add, so she knows our relationship isn't salacious.

They shake hands.

"Your mother said to me about you, 'He might throw a tantrum and hold his breath until he turns blue, and if he does that, you just walk away from him. Don't give in. Leave him in his stroller or wherever he is and walk off till it's over. That's what I do. Don't worry, he'll eventually take a breath, and then stand back, because, oh, the howl that will come out then!' "

"Is that what I did?"

"Not for me, no."

She smiles so sweetly I'm not sure she's telling the truth. Then she holds out her books for me to sign.

The all-around fellow at the Minnesota furnace factory, Dennis Filer, has said, "Call whenever you need to, no matter what. Use our 800 number." I call and ask where the manual is. "In the control box on the right-hand side," he says, and then adds with an encouraging coax to his voice I find calming, as if he's placing a hand on my shoulder, "Don't hire a plumber."

"What do you mean?"

"Ninety percent of our customers install their own, and there's a reason. Plumbers are used to working with pressure heating systems, which ours is not, no not, and they can't get it into their heads that the circulating pumps just circulate heated water from the furnace and back, with no pressure. Every customer who hires a plumber has ended up in grief—unless you know a local guy who will work at your side and do only what you ask. You can manage on your own, right?"

I assume I can, and have no doubt about it if Joseph, who has a gift for mechanics that far exceeds mine, helps. He is like the son in the parable who says he won't and then does, except he doesn't say he won't. He is sometimes slow to start, but once at it he is like his grandfathers: both loved work.

And his concentration once engaged is absolute.

Then my wife reminds me of my commitments, a number of them contractual, and says I must hire a plumber for some of the work, if I want heat before winter. She knows my calculated slowness and is my best business advisor, even if I sometimes resist her advice, anyway at the outset of a project. But not now. We hire a plumber.

I spray fluorescent paint around the bottom edge of the furnace, to mark the spot it will occupy so all connections meet it correctly—the orange glare a shock on the still-green grass. Then I hook a chain to a pair of bolts I've threaded into holes in the furnace's base and with our biggest tractor drag it clear of the fluorescent square. A father-son team arrives with a flatbed of equipment and with a Ditch Witch they dig trenches from the neon rectangle in three directions across the yard (oh, my wife!), to the buildings we intend to heat.

Joseph and I lay pairs of poly hose in the trenches, a supply and return line, inside two-inch-thick sheaths. We make these ourselves from sheets of rigid insulating foam, sawing the sheets lengthwise and assembling three sides with eightpenny nails we shove into the foam with our fingers to hold in place till a high-tech sealer sets, then place the clumsy three-sided contraption (watch the wind!) in the trench around the tubes, add a top, and tape each joint—all to prevent heat loss when the water circulates underground.

I also have the trencher dig down to a pipe that runs close to a three-sided cattle shelter Joseph and the girls and I built three years ago, and once we were done wondered why we hadn't added a fourth side and roof. So in the fall we augered down six more holes, four for the side and one at each end to frame in sliding doors, and now I have the trencher dig down past the line of posts we've set in place and tamped, so when we add the new outside wall we'll have a hydrant inside to use for watering our horses and sheep. All we have to do is add a wall and doors and roof before winter. Which is why my wife worries.

The plumber arrives, and I'll only say that at the end of October, when two buildings were mostly hooked up, the house and the shed where I work, and snow kept falling, Joseph and I finished the rest. It had been a dry fall, except for the September rain when the truck arrived, and then October closed out with rains and snows that came and melted until our yard took on the emerald green of a golf course.

Then on election day the snow started down in earnest. Winds picked up and drifted the snow so badly all roads were grim by noon, and the next day Joseph and I finished the final connections and started filling the water jacket of the furnace in the ingenious way the company devised—by running a hose from our hot-water heater to a return line, so the water arrived in the furnace heated.

This took half a day, because we discovered leaks in the copper in both the house and my office and had to repair them against the resistance of steam rising through each joint from the heat of the torch. Because we were adding water now. Then a pair of pressure-relief valves the plumber installed started spurting water like a baby's penis, and I called Dennis Filer and asked what to do. "Remove them," he said. "All they'll do is continue to leak, because with our system you don't raise enough pressure to close them." A state away I heard him sigh, but he refrained from saying, *I told you so.*

We refill the furnace with the boxes our supplies came in and I light them with our soldering torch. Then I call a friend who was a weatherman in the Air Force for twenty years, serving his last hitch at Colorado Springs, and who is a decade younger but officially retired, an ingenuous boyish presence always about to come to attention, it seems, Jim. He moved to the area with his wife and children to be near his wife's family and the church we attend, and because the brand of predictions he oversaw were based on solar disruptions, he's interested in alternative energy.

He arrives in his four-wheeler, its futuristic topper frosted with snow, just as I'm about to scream at another leak. His yellow-blond hair is raked in a Wildean swag from one side of his head to the other, his glasses of a shape once called "Air Force frames." There is a jittery energy yet an ease about him, as if his inner life is still sheathed in the iron of a uniform. As if to evade my rumbling, red-eyed upset, his sons, six and four, run into my office to check out the computer.

The final leak is in our third-floor room (we have only one), apparent when the water reached its level. Joseph says he is fixing it, *has* it fixed, he calls out a third-story window. So as the snow thickens with a clinging wetness I add scrap wood and branches to the cardboard going in an orange gush like gasoline, then swing the door in when the heat gets so intense I have to jerk my head back from its grab at my face. My eyelashes feel crisped and cling at points and I sense a bunched grip across my eyebrows.

"Crack the door for about ten seconds before you open it and then stay low when you fire it up," Dennis Filer said. Now I know what he meant. When the roar of the cardboard subsides, I add limbs; then logs.

Joseph has joined us and we study a gauge to the right of the door. It measures the temperature of the water surrounding the firebox. The operating range is between 160 and 170 degrees, Filer has said, and a vent on the furnace's side, near the front (a raised square covered by a grille), allows oxygen to enter and keep the burning lively. Behind the grille is a damper door that an electromagnet should slam shut. Or will when a probe extending into the water jacket senses the temperature is 170. When it drops below 160 the same mechanism pops the door open so the fire, smoldering between times, brings the water up to temp again—or this is how the beast reputedly behaves.

We watch for an hour, adding logs, and the gauge only reaches 140—a momentous amount of wood to heat 400 gallons. The November day is darkening at five, although we're a month from our shortest day, and in a fidgety fret Jim says, "I better head for home." The wind is climbing into a whine and this is his first full winter here. We go into the building where I work and find his sons on the computer, one at the keyboard, the other manning its mouse.

I feel a pang of violated privacy when I see that the table beside the desk where my computer sits has tipped, sending a scatteration over the floor, my notes—the "table" actually a scrap of plywood balanced on a brass umbrella stand. One of the boys likely leaned on it, suspecting it would support him, or merely bumped it—this precarious table I've put up with for a year, meant to be temporary. All I have to do is sort the notes, parts of the puzzle pieces to a biography I've been working on for years, and the sorting may solve its riddle.

The three of us who have been outside notice at the same time the warmth in the building, a solid-seeming radiant heat gratifying to absorb. Jim smiles and I pat Joseph's back, then rub between his shoulders.

Jim goes off in his lumbering rig, like an armored car in the softly falling snow, and the slanting flakes draw like a closing drape over it until all that's visible is a rose glow of taillights, then that's gone. And I'm surprised at the vinegary taste of absence on my tongue, that sting of loneliness, so close to abandonment, that children know. We've had few visitors this winter, only our neighbor, Kenny, occasionally, and I've come to look forward to the arrival of UPS and Fedex drivers, so familiar from over a decade of dealing they convey the feel of friends.

In the gloom and dark of snow Joseph and I return to the heat of my office, the interior of which we finished last fall. He is tall and solid, like my father, and now goes to a chair across the room and drops in it as I settle into mine at the computer screen. He has my uncles' languid Slavic ease, as I translate it, though he has hardly met any of them, so that he, like they, can appear at rest even when working—their minds entirely on the task, the areas of their body not involved on vacation, and now he stretches his legs and crosses them at the ankles as he gives along his length like an uncle he never met.

These bonds of blood.

Then his face comes alert with the light of a young person with a congealing idea. "I know one thing that everybody who owns one of these boilers has in common."

"What's that?" I ask. "They all live in the country—away from cities or towns?"

"No."

"They like to be self-sufficient, or imagine they are, or try to be as much as they can?"

"No."

"What, then?"

"They all have singed eyebrows, like you."

In a dark like the dead of night, he and I stand in falling snow, under the furry glare from a 100-watt bulb fastened near the eave of the furnace (so you can load wood in the monster at night), and then the side door closes with a clang so loud I jump. We cheer and slap each other's shoulders. The boiler has shut off as it is bruited to at exactly 170. We run to the house and tromp up the steps to the upstairs, where we've installed radiant baseboards.

This story was always frigid no matter how much heat we poured on and now it's warmer than downstairs. The ground floor is fine—a heat exchanger in the plenum of our gas furnace, which now uses only its fan—and from there we go through the mud-room at the rear of the house, which you enter by an outside door in order to wash up, and down the back stairs to the basement. The runs of copper in this damp and clammy place, with its exposed earth and walls of red-orange tile eroded by the webwork and crystals of nitre Poe draped through his stories, have raised its temperature a dozen degrees.

"Wow!" I say, hearing a sixties taint. "Warmth!"

Only one building left to hook up, but it means running copper the length of another—one we might also eventually heat, a garage-granary. Though it's pitch dark it's not time for dinner, so Joseph and I carry tools and copper fittings out in buckets to begin the run. Snow sifts under the garage door and past its edges, although we've installed new seals all around, and when we can't find an exact elbow—a street 60—for the first fitting, I say, "Let's quit for now."

3

I wake thinking of our daughter Newlyn. She was our only child for nine years, so close to me and my wife when were young, still developing—as it seems in the misty memory of half sleep—it's difficult to set her at arm's length, as it were, for a clear look. She's an essential extension of me, separate yet not, sending up heat like a second conscience. I can't make out why this is except I recognize she's upset, her face that's her mother's drawn by a gravity of emotional weight like grief.

I drift to the year she was twelve; was five, was two, her calm but ardent nature, with a risk in her look that catches the attention of others: its unblinking openness, the tie to what endures between my wife and me; and she walks or runs from one to the other as if to perfect a weave she senses or to draw it tighter, a dutiful act of a dutiful, only and oldest child.

The fall when she's three my wife returns from the store with her and starts to put a package of corn on the cob into the refrigerator when her arms fly out from her snowsuit and she cries, "No! No! No! *Corno!*"

My wife usually summons gentle concern for others and now she asks what it is our daughter wants and tries to understand, then closes the refrigerator door, its pneumatic thunk the end of things.

Newlyn pitches over backward in her snowsuit, kicking the floor with her heels, wailing, her mouth wide, and my wife turns to me, astonished, as if I'm the cause.

And then (I'm ashamed to say) we laugh, or anyway smile at this overwrought reaction entirely new in our daughter, while she keeps wailing. Bulblike tears of hurt spill into and over the blond curls in her hood. My wife goes down on her haunches in a mother's move of rescue, and with that the entire scene turns yellow—the corn on the cob, my wife's yellow-blond hair, Newlyn's snowsuit and curls, and then the infant in me holds his breath in the stroller until blue tips streak across the yellow, and Newlyn comes climbing through the blaze in a heated clamor as my heart knocks, her knocking for entry, she enters, the accelerating thud of it, then the airy bubbles of the oxygen of life continuing in their run.

We were that close. And much as the event may be an embarrassment to each of us (to me for my amusement), I know this is the moment that sets me on course toward a decision that will take several years to seal off: that my wife and daughter are as important to me, or more so, than any career or ultimate employment, by which I mean writing. But I will make sure my writing is as good as it can be whatever genre I might use in whatever phase I'm working through, so that the best patches will lodge in a consciousness like a coat on a hook.

And though I might be far inside the house with one or another of my family and may indeed never return to fill the drape of the coat again, there it will hang for good.

Or so the thought clings as I slide into sleep.

Then I'm jolted awake at the picture of her baptism, in the vaulted baptismal sanctum at the Church of St. Vincent Ferrer. As the priest applied oil (or was it salt?) to her forehead and said, approximately, "I hereby cast out Satan and all demons," she erupted with rattling gas in her disposable diaper, and the rattle echoed in the high-ceilinged sanctum. If only evil were so easily dismissed.

Both my fathers, as I viewed them, my blood father and my literary father, Bill Maxwell, were present. Bill's elegantly handsome wife Emmy was godmother to my father's godfather, and when my father died the Maxwells took on with more seriousness than most their role as godparents; they continue to write my daughter and send her gifts.

After the ceremony, we stepped into brilliant sunlight on Lex, and I invited them to dinner at Paul Revere's, a restaurant known for its steaks and yard-long steins of beer, set in pine holders at your table—the only single beer that satisfied me. We were driving a 1963 Bonneville convertible, one of the longest cars built, bar none, and my wife nestled our premature daughter safely between the seats in a picnic basket, the simplest way to carry her with full support. Since I was driving us all to the restaurant, things had to be shuffled a bit, and I believe it was then that Emmy Maxwell first held Newlyn for any length of time. "No squeaking!" she said, whenever Newlyn's pure and high-pitched voice rose in a protesting sound she produced.

After the meal at Paul Revere's, as we stand at the car saying good-bye, I say to my wife, "Wasn't the camera bag in back?" Meaning on the tonneau flap inside the convertible's zippered rear window; it had been. I unlock the car and see that the bag and our Polaroid camera, with shots of our daughter and my father, besides all the shots with the Maxwells, before and after the baptism, are gone. "Why didn't they just take the camera?" I cry. "They could have had it! Why the pictures?"

Bill Maxwell slips both hands into his trouser pockets, as he does when he's troubled or about to make a momentous decision he would rather not, his suit jacket flaring over his wrists, and stands in silence, eyebrows raised so high in concern his forehead corrugates in wrinkles.

"If they had simply taken it," my father says, "the police might think they'd stolen it, but wouldn't think that if somebody had it in a bag with all the rest."

"Goodness!" Maxwell says, entirely at ease when he laughs as he does now. "You sound as if you've lived here all your life!"

My father glances at the restaurant, disgusted by the beers I've had, as he's already told me, and says, "I figure this one day here is about enough for me."

Maxwell laughs with a delight I imagine I had as a child.

The next day, with the storm ratcheting up, our electricity goes out. This lasts an hour as I fret about the pumps connected to the lines leading to the furnace (now idle) and the temperature of the water in the beast itself. It takes electricity to open the damper and we don't have the inverters that Dennis Filer recommended we should buy.

"You hook one to a twelve-volt battery," he said, "and it runs a pump for a day or two—the damper-door switch for who knows how long. I had just one pump going like that for a day and it kept my whole system warm with the seep."

I flounder through deep snow to the garage-granary. Our tools are below a drift inside the building that has already reached the top of the five-gallon buckets of fittings we carried out. It's filling them up.

This can't go on, I think, and a chill like a shock leaps from my knees out my shoulders.

The next day, the electricity goes out again. We wait an hour without a hint of its return, not even the usual occasional flicker, and finally I call the company, a rural electric co-op. I explain our dilemma, hundreds of gallons of water in a furnace that could freeze, not to mention the underground runs and the copper in the house and buildings.

"We know a line's down south of you," the person to whom I'm trying to explain our situation says, "but we aren't sending anybody out. Our lineman tried to leave town early this morning to fix another outage and had to turn back. They couldn't see three feet out the windshield. The storm's that bad here."

I ask when they expect to have it fixed.

"I don't know. We aren't thinking of sending our men out till this stops. We can't risk their lives in weather like this!"

I sense a tone of accusation and as I hang up I think, Aren't they paid for the risk? With electricity, yes, when a line is down that could kill or with transformers and relays carrying voltage capable or arcing ten feet if you aren't properly insulated or slip up, yes, I see. They can't add to that the danger of possible death in a storm that will let up, any more than I would feed my horses and sheep (which I have to feed) if it means imperiling my life or my family. Right?

So I go to the front window in the dim and cooling house and look out on a horizontal blur of blue-white. In such weather all you can do is look out a window, if it isn't frosted too thick to see through, and try to gauge any letup. Above or through the action of the snow a steel blue pall of partial light seeps, though it's noonday, then blank blue-white, from blowing snow that appears to fill the universe.

From the trunk in the attic I take out and shuffle a deck of memories printed by the years in Sykeston (about as thick as the pack accumulated over the following fifty years), flipping past the ones I've given away or patched to characters who have grown apart from me or gone off on their own with them, and realize I'm looking for a winter to match this one.

Then it comes, the winter of 1949–50, our last in Sykeston as a family, in every sense, and I find myself in a narrow walkway dug through drifts over my head. My father's work. He has pioneered the way with a scoop. My numb feet clump over gritty crystals that slide, then slivers of bright green glint. Grass in the ice. The sun feels far off and feeds an icy-blue smell of crystalline layers at my elbows. A hint of grass crushed.

Smell is my primary guide and it is smell that leads me to the heart of a story—that animal sense that children especially possess—although the best of the gifts I've received seem visual. I have to see it to write it and then see it to understand what I've written.

The path cut through snow is so deep my brother Dan and I have to kick holes in its sides to climb to its top, then paw with our mittens for a place to hang on. We put a piece of cardboard over one stretch, piling snow on it, and our dad can walk underneath it if he stoops. I see his grin at this accomplishment and think of how he addresses us in a teasing or joking way, as if he can't handle us as he does his high school students and so holds himself at a distance—unless we disobey. Then he's angry and direct. He later tells the story to a friend of how Danny and I made a roof over his path and his telling prints our actions over the snow in a scene as vivid as the queen of hearts. Do I remember only what he said or added to?

Or her. I follow her pattern, refined as a dance, from the stove to the table to the icebox to the sink to the table to the icebox to the stove, and see that her walk has a swaying beat, as the songs she sings or presses from piano keys, tapped out by the heels of her low brown shoes, and then she turns to where I sit at the table, her eyes going blank and wide, and says, "Just who do you think you're staring at, Mister?"

Besides those two and the blue and green (now gray and white) it is the surface of the earth I retain: the polish on tin that bears an impression of bricks—a high seat where my father sits me, the top of a housing like a low shed over a mechanical motivation I feel through my butt when he throws a switch that sends the yoke of a pump leaping to life. It rises and falls, dropping to its knees like a sinner after confession, then jerks straight up, *hello?*—the town pump, our source of drinking water. Then the airborne descent with my weight suspended at my armpits, the impact of my feet on ice, his gloved hand closing over my mitten and the long cylindrical pail with its fitted cap in his other hand swinging like a pendulum in time to his steps.

Or beige bricks with lines like flesh cuts down their fronts. Over them wavy reinforcing wires, crisscrossing one another in diamond shapes, form protective panels over the ground-story windows of our grade school. I place my tongue on one in zero weather and the disengaging rip leaves a burn worse than the scald of cocoa for days.

Sister Benedict is at her desk set at an angle across the corner of the primary room, smiling, her wimple with its white headband like Father Sommerfeld's clerical collar, her white circular bib shiny with starch over long black clothes.

Father stands in a black overcoat and black homburg near her desk at the front of the room, sheets of sun enameling our desks, the green-burned smell of cigars clinging to his coat, and speaks like an actor as he reads off the grades from each report card before he hands it out, after a pause of taking time to look up and down through his gold-rimmed glasses at us all, a way of putting us in place: "*Rudy!* A C in cheogrraphy? C in singing means like a ganary but in cheogrraphy? *No!*"

The wind groans and drifts rise two feet up the sides of the furnace, none of it is melting. That means the furnace must be insulated the way the manufacturer maintains. The steel door, though, radiates so much heat it forms an angled drop-off in the snow, turning icy from the melt, so that I slip down it or logs slide down, as one does when I slip, denting the metal panel covering its front. We have little wood and even if we had wood there is no way to regulate the air without electricity.

Then I remember Dennis Filer's story, when he was recommending an inverter—how, before he bought one, he had to prop his draft door open once when the power went out. I remove the grille, draw back the door, and set a wedge of wood below it. But we can't circulate heated water from the boiler to the buildings. No juice. I hope for heat seep.

The caragana around the house in Sykeston sets a boundary that triggers our sense of freedom: the beyond. Some boundaries are for our good, I learn when I fall from a tree I was told not to climb, and some are meant for our well-being, like milk over the lumpy cooked cereal I dislike, and what might be most important and should head the list: the melt of a communion host on my tongue, its sticky flat hole opening to my toes. How, after that, I'm new.

Other boundaries are to break. The first is age, whether six or eight—along with the limits at each level, because every level has a few—and next what we can't do because the neighbors don't want us to. Sometimes they don't care, I find, when I walk in a garden where my mother told me not to, or knock on a door and ask for candy, or fight with my brother where neighbors can see. But other barriers are invisible.

If you feel you shouldn't enter a yard, don't. A mean man or a dog that bites will arrive. The fence of wavy wire around a yard, its pattern of arches repeated in loops, is not to keep you away but welcome you, because the friendliest family in town is waiting inside.

An openness like a field beyond a house, or the *feeling* of a field, is a place I do not go, because no field is there. If I walk past our hedge, over the sidewalk, through the ditch, to the edge of the street and look left I can see one end of the main street, and in the other direction blue and green. It gives out in a row of posts that are telephone poles.

Down the cross street that runs in front of our house is a misty-blue elevation, the farthest run of a line of hills that starts at Jamestown and follows our car all the way to this end, Hawk's Nest. Indians surrounded the Cavalry there to starve them out, but the soldiers had horses and somebody sneaked down for water and got it back to everybody on top. Dad told us. He has the tall and perpendicular forehead of the man in many books, Shakespeare, his handsome wavy hair combed back.

On our farm I sense boundaries I'm disinclined to approach, such as the tree to the left of the sidewalk where, when he was four, Joseph tipped over a can of gas. I spanked him for the waste with an anger out of proportion to the loss or harm to the tree. And three steps to the other side of the walk, I laid a storm window flat and he stepped through it with bare feet and cut his big toe badly. Then a pace ahead is where, at four or five again, he wouldn't admit he had smashed the brass tube of a hand sprayer and I put my hands on his chest and shoved him so hard he went stumbling backward and fell.

Of course he hadn't smashed it, as I learned.

So at this spot I tread softly or hurry past, feeling a stir of sensation over my shoulders, unable to forgive myself for any of what happened here—though I've asked my son for forgiveness and received it, I believe, at the time of the incident and when he was an adult and could remember only stepping through the glass. "And that was my fault," he said. "I shouldn't have been there. I think I did it on purpose."

And I feel again I'm behind a zero-cold mesh over a window, reaching with words on my tongue toward the people in the misty scenes of my story below.

Handling Too Many Candles

I set thermometers in every critical spot, beside the baseboard heaters upstairs and in my office and the lowest copper in the basement, and start rounds to keep watch. Every four hours—as by experiment Joseph and I learn—we have to drain water from the furnace into the basement of the house and the greenhouse attached to my office to keep the buried lines from freezing underground.

Now and then I remove the prop from the damper door, when the water in the furnace starts boiling, because this can damage or impair the furnace, as Dennis Filer has warned. It isn't a steam furnace, it doesn't work under pressure, it's not meant to boil, he kept saying, and the only way to move water out of it, into the pipes now residing in a kind of permafrost, is by electric, circulating motors in each building. So of course the more we draw from it when we open drains in the buildings, to warm the lines underground, the more it tends to boil. I also listen for a boil on my rounds.

After it has been dark for hours and my feet feel as cold in the house as outdoors, I turn to Joseph, who has started following me with concern, and say, "We can only deal with one building. We'll have to drain my office."

The plumber didn't install a drain at the lowest point in the copper, near the concrete floor dropped four feet below ground level in the greenhouse, so we have to cut through the line of copper there by flashlight, install a drain, then enter the storm and take a side panel from the base of the furnace (after digging down to it) and under the furnace shut off the supply and return lines to my building. Then we let the crystalline cold water of that loop drain into a sump we provided when we poured the cement.

"There are ninety-degree drops and curves in all the copper under that floor," Joseph says, of the building itself. "What if it all doesn't drain out?"

"We need the air compressor we rented to test the lines once we got them in," I say. "If we had electricity to run it, if we could get out of the yard to get it."

"Right."

"If we don't get all the water out, some of the copper is going to burst. Let's bring the electric heaters out," I say, regretting this return to a source of heat I hoped to eliminate. "Maybe they'll put out enough heat to keep the place above freezing."

He looks at me as if I'm angling for another joke, when he's probably had enough of everything for the day.

"Right," I say, though I meant what I said before his look made me stop: *no electricity*.

We get a kerosene heater going and I set it close to the baseboard heater that extends the length of the north wall, under the desk where I work. I get our lantern flashlight and shine it at the gauge on the furnace, then have to step closer to read it through the snow, that thick. 120. In the house I see my breath in the wobbly light from candles my wife has going and I stop at a floor-to-ceiling bookshelf past the entry and turn to the substantial trailing shadow a head above me. "You better get some rest, Joseph. I'll stay up."

"No, I will, Dad."

I raise the beam enough to see no resignation, his head high, eyes clear. He is the craftsman, the adept at the actual, I the journeyman of the interior, trying to find metaphors for the actuality he modifies, and for the last weeks the final plumbing and repairs and engineering of it all, I realize, has fallen on him—a nineteen-year-old who likes his sleep as much as the next.

"Thank you, honey," I say, and crimp my lips at the slip, this endearment I've used since we began to talk as two, though he doesn't flinch as he did a year ago, as if our predicament warrants every shard of solidarity.

"I better," I add, setting my authority between him and any guilt he might feel about sleep. He is that sensitive, I can say now, although I was mostly unconscious of it—his own father!—until my friend, a Japanese composer attuned to the nuance of underground chords and family ties said, "He's the most sensitive boy I've seen."

"Thanks," Joseph says, sensing at least partly the movement as of a primordial sea swaying under this moment.

"God be with you," I say, my nostrum.

"You also."

"Sleep well."

"I will," he says, and turns away.

"Thank you for your help." And then in haste, as if under cover of the dark, "I love you."

He pauses as if stopped by this, then looks over his shoulder, straight at me. "I know," he says.

I listen to how his weight wrings a response from the wood and nails of each tread, so different from the steps of anybody else, and think, How light his walk once was!

I blow out the candles, so many they seem my birthday cake—*too* many, I think, with my wife in bed, asleep. I sit on the couch in my outdoor outfit and shine the lantern at the picture window. She's hung a dark quilt over it like crepe. I shut off the lantern and drop it in my lap. The beaten hollow of death stalks all. I feel its claws testing the hair at the back of my neck for ripeness, as I have all winter, though not as my neighbor, Kenny, must.

I profess to be a person of faith but most days I'm numb to the supernatural. I would expect it to exist, at least in a perfect sense, simultaneous to the world we walk through and touch. I tend not to trust my intuitions about the presence of spiritual beings, even when a force like a microwave signal fires through me, or when I sense my mother or father. Nor do I trust myself, wholly, about the other side, those manifestations over my fouled patch of ground, or the ground crawlers, wall huggers, door swingers, sources of unearthly noise pitched so high an eardrum depresses and then pops in release.

But I know they are not to be belittled or treated as less than ministers of a power that mars the earth and every form of life on it, including the most marring, human beings. I'm aware of the evasions I use to dismiss any acknowledgment of evil, even though I know that is the exact modus operandi and rationale of evil: *Hey, I don't exist.*

Evil is so easygoing it has you before you know it and nothing is wrong about what you do. The Russian exile and poet Joseph Brodsky wrote, "Such is the structure of life that what we regard as Evil is capable of a fairly ubiquitous presence if only because it tends to appear in the guise of good. You never see it crossing your threshold announcing itself: 'Hi, I'm Evil!' . . . A prudent thing to do, therefore, would be to subject your notions of good to the closest possible scrutiny."

Brodsky points to the fallacious use of a teaching of Jesus by another Russian, Count Lev Tolstoy—Gandhi's mentor—in the stance we now know as "passive resistance," and offers the 1984 graduating class at Smith College (where his remarks were first given as a commencement address) the antidote: You can overcome evil only with good.

The real claws are at Kenny, our neighbor. Two winters ago it was his wife, *Valeria*, a cousin in pronunciation to valerian, the tranquilizing herbal—our dark-haired Dolly Quickly, always in such a rush the backwash of dust from her car formed a contrail, she its bullet head.

My ears open to the night. Only wind. Silence is our selling feature. No sirens slicing a sentence at its center. We hear every car or, more often, pickup, for miles, depending on the wind, and a deeper rumble rose from Valeria's contrail as she slowed near our place for the children. Then we paused to see if she would turn up our drive and dash from her car to the house to drop off carrots or küchen or rolls warm from baking, in and out so fast I asked if the blur was her, trailing her phrase "Praise God!"

Sixty, maybe, but with the energy and trim of a woman of twenty, a bawdy devotee of summer sun who gardened naked. And played piano in a bar band from the age of sixteen until she had children, a daughter and a son, in that order, both extraordinary in her eyes, because of operations for "female troubles," as Kenny put it, she thought she wouldn't survive. She took correspondence courses in theology and in her last decade embarked on as much as she could do, first traveling the U.S. and Canada, then Europe on gourmet and music tours, then Caribbean cruises with Kenny, when she was done for another year operating tractors and heavy machinery. She also kept a garden with a French purity of design—her time to indulge herself in the sun—and when Joseph and I built an addition on the side of the house facing her direction, south, setting one room on a story as tall as the wind-charger tower—she stopped by one day and looked level at me and said in her husky alto, "I hope you don't plan to put a telescope up there."

Then her jetlike run to town to shop or visit her dozen friends, the sort of woman others competed for or hated, often to pray with them, or with a backwash in her rush slowing to visit her parents and, after her father's death, staying at her mother's longer, overnight at times—Valeria, this "friendly little lady," as my mother was reported to be, who was working her way into me as a mother—now gone, a cancer of her pancreas taking her so fast she was dead in a month, lying in the casket composed, a faint smile on her face, all that energy so monumentally still I try to breathe for her.

I look from her to her son, his blond ponytail sweeping the back of a European-tailored suit as he turns to somebody and nods. His hands are clasped below his waist and I go and take one in mine and say, "There was no one like her, your mother—an extraordinary woman, a true saint."

"Oh, I know, I loved her so much," he says, and smiles her unabashed smile. "She didn't want us to mourn, she wanted us to rejoice that she's with the Lord, but I miss her." He takes my wife's hand, and I think that being a saint is not predicated on who you are but what you do before anybody asks.

Sleep starts its numbness at the back of my skull, behind each ear, and I try to relax in a fashion other than the way of the one that leads to sleep, a discipline a writer learns.

And what happens in sleep, in that deadfall from a semblance of control into oblivion, entering blackness in the final drop?

Often at night, as I slip from the shore of the everyday, faces I've never seen enter the edge of sleep, illuminated with a supernatural clarity. They are speaking, carrying on conversations about their own concerns, although they are pulled in place, it seems, by an inner eye that intensifies in focus, examining them with the awe of objectivity. One face will talk with easy familiarity to another—a switch to it—and the familiarity wakes me quicker than the swim of either. Or several will speak together with the politesse of people onstage—talk that travels through so many convolutions the story is hard to catch. And on occasion, at the edge of sleep, the voices take on reverberations that seem engineered—a hallucinatory echo rebounding across the night.

If I wake, all's a blank, not a word remaining, though after half wakings I can sometimes edge into a scene or draw it near in partial sleep. As last night, when a family of Africans in casual clothes sat on this couch and spoke with such concern for me I found myself awake in bed, the thud of my heart setting off quivering springs, with only a glow of their faces fading in the distance, and one word: Anteropia.

Not Ethiopia, I had to tell myself. *Anteropia*.

All of this, as I understand it (and I understand it only faintly), feeds whatever builds a pressure of words and inclines me to write. I listen to these people and others, real people, usually, but the replaying is in private: *born near winter, how you look forward to it, that still time when a phrase assumes a molten perfection. Seal it in winter, a season barely lived through, the worst you've seen.*

2

I'm six, in my first year of school, helping prepare to go to Minnesota, to Grandma and Grandpa Johnston's for Thanksgiving. Our father packs the trunk and our mother brings out more. He rearranges everything and then takes it out and packs it again, except for a carton he has to set in the space between the front and back seats, where Dan and I prefer to ride. My mother carries out another box.

"Audrey, what's in this one?"

"Clothes they can use and things I've baked and some vegetables from the garden." She turns to the house, then back. "I have to get a few canned things in another box."

"But, dear, so much?"

"They are destitute, Everett, and need every stick of help we can haul them!"

"I doubt they're destitute, dear."

"*Poor*."

"All right. But where do we put another box if we're all to go?"

By the time we leave the well between the seats is filled, so Dan and I can't hide there or even let our legs dangle. They stick straight out over boxes. And on the way my mother buys gifts and finally lets our father buy for Grandpa what he wants, six brown bottles of malt, after he has explained, "The malt has no alcoholic content, dear."

"If there's alcohol in it and he gets going Mother might as well try and stop a train."

From the roof of the shed where Joseph and I work we see Kenny turn up the drive in his red pickup with its amber twirler on top. He has set aside half his farm as a hunting habitat, two sections of grass, and he patrols the roads in his pickup to keep the yahoos in line. This is pheasant country.

The season is supposed to last to the first of the year but recently all we've seen are rare suburban hunters, as I call them, because they migrate from the suburbs on a weekend in pairs, usually in a vehicle of that name, and cruise down the shoulder of the road so one can get a leg out and fire quick (even out an open window) at pheasants rising from the ditch before they reach the height to head in the plumb-line of their dive, out of range in seconds, out of sight.

Kenny gets from his pickup and walks our way, hands in his pockets, without his usual animation, fists out, bare, swinging to his quick walk. I want to head down the ladder but also want to finish our roof before the winter gets worse, and we're setting the final rafter in place.

"Are you guys going to get that done before Christmas?" he calls up, as if he's read me out, a big man like my father, six feet and built bearlike, with black hair like my father's combed back the way my father combed his, the bill of his cap tipped above a widow's peak.

He watches as I help hold the rafter in place, the hardest to handle, at the edge of the building, and once Joseph gets a nail in his end I grab a scrap of wood and tack it between that rafter and the next one back and go down the ladder, leaving Joseph to finish up.

"It's Brad," Kenny says. Joseph stands, his head above the harp strings of rafters. "He was at a show in Paris—France," he adds, in case we wonder. "He had to set up an exhibit for Sony and it went fine, he said, but some of the equipment was so heavy when he went to pack it up he thought he hurt his back." Brad, who is with Sony New York, first went to Japan as a missionary, married a native, and learned the language so well he is the liaison between Sony America (along with other duties) and Sony's hierarchy in Japan.

"Then he had to set it up again back here, someplace in New York state, and said he started to lift a speaker and felt like he broke his back. He went to a chiropractor who sent him to a doctor who said, 'Why didn't you come sooner?' Why, he had the yearly exam Sony requires six months ago! Why didn't they catch this then? These damn doctors, by God, excuse me, but it's as bad as with Valeria, because if you opened the door on any hospital during duck-hunting season they'd all be dead, *Quack! Quack!*

"So this doctor says it's cancer. He took Brad in for treatment and now he's so weak, he told me—he just called—he has to walk with a cane. He's not even forty-five!" Kenny looks over the white sea of our fields, drifts buckling over others, and I see the winter Atlantic, gray-green, with its brocade of capillarylike foam gliding over rollers growing into breakers on their way in.

"I'm sorry, Kenny." I turn to Joseph, whose look is his mother's, *empathy*; it was hardly a year ago that Brad and his wife Kiyoko had the child Valeria hoped for years they would, a daughter named Maya. "How bad is it?"

"They told him it's so advanced in his spine they can't hold out hope. But he's not giving in. He started an alternative treatment, too. They give him something orally, then wrap him in wet sheets so hot he can hardly stand it, and he said the sheets come away brown from the poison. 'I have Kiyoko and Maya to live for,' he said. 'I want to celebrate Maya's first Christmas.' "

Kenny grips the bridge of his nose as my father did to stanch his grief when my mother died, and I put a hand on his shoulder. "I'm sorry," I say. "We'll pray for him."

"Do." He stalks away to his pickup.

All those flowers, no. The reek of so many varieties in a funeral parlor could kill a bee. As at my mother's funeral. When she died the years in North Dakota tumbled into the attic, because we had moved to Illinois the summer before, and her death was like a guillotine across them. They went falling in folds into their place in the trunk and its lid slammed shut.

Her grave was a chasm, the spot where the guillotine struck, and when I tried to get to the other side by every imaginative leap I could devise, it didn't matter if I missed by a mile or an inch, I was in a dark deeper than dreams. Even after I made pathways to the trunk and dug deeper in it each time, no way was safe back to adulthood. Most of my work rests on what I learned in that seven-year span. This side of it, only a year from the winter my brother and I placed cardboard over the tunnel Dad dug, was the fault spot where the guillotine struck—separate worlds I use words to unify.

At the worst times that first year I could drift back to our yard in Sykeston, the grass crackling under my knees in newness, enclosed by caragana with its shiny coppery bark approximating snakeskin, its blossoms like pinpoints of heat from the sun, and rest within a single scent below the grasping twirl of bees on yellow bells—the world that sustained me through her death. Just last January.

No, Valeria—these awful leaps that divide me when somebody dear to me dies, so that in defense I go rigid, frozen from sympathy, in a terror of self-protection.

I snap on the lantern and go to the window as if to see out. Through the quilt and the double panes I feel a radiant chill from outside. I swing the lantern beam aside and picture the earlier, quiet nights of winter, when I would walk out and see a cascade of Northern Lights, or that phenomenon caused by moisture in the air—the headlights of cars on the highway shining straight up for hundreds of feet, like huge pencils in a hand composing lines of prose as long as the highway, as long as the history of the race, oh, Eve.

I make a round of my thermometers and see that all have dropped. Was I asleep? Outdoors, hearing a boil in the furnace, I pull the prop. The wind is worse but when I slog through swirling drifts to the back door and go down to the basement in creaking steps, it feels warm. In contrast, perhaps, because my breath billows, and when I start to drain water from the underground line it's so cold I'm shocked.

All this with a furnace we haven't had a month.

"All right," I say to myself, in practice for the tone to use with Joseph when I wake him, "I guess we're going to have to shut it down. I'll need your help to carry all that water out." *Four hundred gallons*. Maybe down to 250 now, I think, and at that second a noisy flash sends my arms over my face in protection. I see a glow and hear the furnace fan running but it takes a while to realize the electricity is on, and like Valeria I cry, "Praise God!"

I run upstairs and shut off the lights left on when this began, then hear steps start down from the second story. Joseph appears.

"It's on!" I say.

"I see."

"Now we can get everything going. Let's see, first we should get the furnace up to heat and—Let's see!" I'm seven, seeing a shining bike for my birthday, and I take hold of my head with both hands as if to hold it in place. "Or maybe we should wait a while and make sure it's on for good. We have to fill the furnace, but first we should get it up to heat. Then we have to check the lines in my shed. Maybe we should do that now, since we've got light, before they get worse. The electric heaters will work! And the pump! We'll have water pressure to fill the furnace and wash up and the rest!"

We've been using the blizzarding outdoors.

I head for my office, as I've suggested, and notice light trying to crowd at angles through the snow. It's morning. But the cold, its unchanging presence, is the antithesis of light, color, the verdancy of green or the red of running blood, the clamor of life. Only a shattering stillness my hearing strives to fill and finds a wind so violent I wish it would shut it down.

I hurry into the building where I work and walk into an atmosphere of dead cold. The kerosene heater is out.

There is a time when a father feels only tenderness for a daughter, then stormier days arrive. They arrived with Newlyn, then somewhat with Ruth, but not Laurel, not yet. Since she was unexpected and surely our last, we not only named her Laurel ("The only way I'll be crowned with it," I said) but gave her my initials, the closest we came to naming a child after either of us, using as middle name my wife's paternal grandmother's maiden name, Andreson.

Laurel was crown to the child-bearing days of my wife, that fruitfulness, a crown to our marriage, now that we were wed with the greater unity of years, and the day she was born it struck me that when she was twenty I would be sixty. Then I remembered Bill Maxwell saying to me after Newlyn's birth, "The next thing you know she'll be a teenager and you'll be an old man."

And, oh, how he laughed, as if he'd shake out of the chair at his desk, laughing as he did at statements, like this one, that got to his funnybone so fully it didn't seem they'd let go.

3

Laurel is in the car with me, alone, one of the few occasions I remember being alone with her (she's usually with her mother, now visiting *her* mother), on the way to Bismarck, to a chapel inside a care center, where I've been asked to exhort—a Presbyterian euphemism for what happens when a person who is not a pastor or "teaching elder" is asked to speak.

We're early, so Laurel and I set up chairs in a room that has the feel of a funeral parlor, and the floating fear I feel when I'm the one who looks for others at the Word balloons under my stomach. Then a young couple with four children, a number I'm used to, walks in, the first to arrive.

The handsome woman cradles a child in one arm and leads another, her husband twins, and before they reach us the woman is talking about an article in an airlines magazine, about the Badlands. They had to fly East for a family funeral, she says, and pulled a magazine from a seat pocket while in flight, and out of the blue (she smiles) they found my article about the Badlands of North Dakota—the coincidence of that, considering their journey, helped settle them, she says—now just back. "And here you are."

"I believe so."

They are attentive, gracious, the kind of couple that people in the city would call, sans children, "beautiful." But it's hard to receive what they say through my state, and finally I turn, sensing my daughter at my back, a close follower, my height already at fifteen years.

I say, "Have you met Laurel?"

"No."

The genial young man in a double-breasted suit of the best cut, fitted to his squared-off shoulders and chest of a weight lifter, reaches out to shake. Laurel, in a reserved version of her several social smiles, scarcely shows her teeth, then gingerly extends a hand.

"No, we haven't," he adds in amplification, taking her hand in both of his as he studies her. "But I sure see she's your daughter. Those are your eyes."

An odd way to put it, I think. Then Laurel and I turn, our heads at a questioning angle, as if we're looking over a windowsill from opposite sides or, better, in a mirror. Nobody has mentioned the similarity and our eyes jerk to take this in, then hers soften, the bright blue of the sky, and I feel a salt sting in mine with a sudden intuition, a glimpse, of how hers will run when she looks at me in a casket—a weighted intuition that arrives with a shock.

No pity or morbid introspection but a depth of coolness as of death itself, and then I want to comfort her in her state I never would have imagined. She returns my stare (politely smiling) as if to view her eyes, and a delicate tenderness registers, the essence of her, a first full inkling of it; she often keeps others at a distance with her ripostes and tart comebacks, breathtaking in their cutting precision since she was six.

But that has been giving way, as I've mentioned to her in recent talks, hoping to draw out this new nature in her: this dutiful, gentle tenderness I've had intimations of but haven't registered with such depth.

My fault, I think, turning to the couple, because I withdrew from her when she was five, in reaction. My wife was spending hours with her in her bedroom at night, reading and holding her in her lap, as if our last child were a lifeline to a better self.

"Yes," I say to what I've heard, confused, caught in my cold-hearted withdrawal from Laurel—so intelligent that as a two-year-old she had to have books in her crib to fall asleep, and how this manifested itself even physically when she was four and I cut her hair, thick and heavy as hemp, and saw it draw up all over her head in ringlets, springy curls (not one strand of that now), as if the dimensions of her intelligence had been

freed and now—oh, Lord!—register as tender admiration, this lovely Syrophonecian woman willing to be happy with mere crumbs of affection that fall from me out of a love that in my carelessness I've all but turned aside.

I shake my head to clear it. I have to, ah, *exhort*.

As I do I observe her attentiveness and understand the hovering brightness in her eyes, and then, for what feels a first in decades, I stand before a group humbled. What I talk about is from notes and doesn't touch on what has happened except for the tremor of feeling filling to the brim sentences and phrases that were mere arrangements of words before the young man (staring at me so intently he appears to know what he's done) said to us, "Those are your eyes."

"One Christmas was so much like another, in those years, around the sea-town corner now, and out of all sound except *the distant speaking of the voices I sometimes hear a moment before sleep*, that I can never remember whether it snowed for six days and six nights when I was twelve or whether it snowed for twelve days and twelve nights when I was six."

Nor can I (though this is Dylan Thomas, the emphasis on the voices mine) as our awful winter compresses and shifts like the snow itself. When I was a learner like Laurel I listened to Thomas on Caedmon records in college, and his rollicking language, as if he cared not a whit what others thought and so was free to be outrageous—the swamp and welter of expression—came as a revelation to one so solemn, close-mouthed, and cautious as I, with the reverence yet mistrust for language of those who grow up believing in prayer, then see a parent die young.

I missed Thomas at the university I attended by years. His sitter for the night, a faculty member who felt he had to assume that role, claimed Thomas jumped up and down on the bed in his room, naked, crying, "I want a woman!" I missed him but saw Robert Frost.

Poetry readings were the rock concerts of the day, with spotlights haloing a poet's head like the nimbus of public regard, and Frost's reading was in an auditorium that held a thousand and was packed. I was up front; I had waited in line to guarantee that. He walked onstage to a sea-wash of applause, looking pleased, nodding, his white hair agreeably mussed as it was during his televised performance at John F. Kennedy's inauguration, and finally quieted the audience.

"I'll say some short poems first," he said in a jerky baritone. "It's a way of taking a reading of you." In case we missed his meaning, he added, "So I know the level you're on."

He cried, "It takes a crack of the quip to make Pegasus prance!" Then he stared over the audience with a majesty the most self-important professor could never summon. He declared another and cupped one hand behind a hoary ear that looked the size of his hand.

He said, "We dance round in a ring and suppose,/ But the secret sits inside and knows." Scattered applause, then an ovation, and I figured it was a code for the atomic bomb, which everybody present had a fear of being fried by, and he sighed "OK" and started reading. He knew every poem by heart and hardly finished one before people shouted "Birches!" He pretended not to hear or gave a stony glare but when the reading was over and he shuffled out for his second encore, with everybody shouting the title, including me, he gave in.

I used a front exit to get into the wings and found a crowd gathered around him—he was beside scenery ropes, under a naked bulb—holding up books for him to autograph. I merely wanted to shake his hand. My only acquaintance with poets was out of the pages of anthologies so I had trouble imagining they went to the bathroom. When it was my turn and we shook, he held out his other hand for a book. All I had was a folder of poems from high school that I brought to pass the time while waiting in line. I pulled one out, reversed it, and he bent to a table he was using for signing, then swung and said, "This is a poem." He had turned the page over.

"It's an old one, pretty bad—all the paper I have."

"You aren't going to take this with my signature on it and sell it for the money?"

"Oh, no! I'm going to cut it out and paste it in a book of yours. I've been reading your poetry for years!"

He pointed to my name at the bottom of the page and said, "Am I going to be hearing from this guy?"

"How?"

"One line is fair. But if you try to sell this and I hear about it, I'll come after you."

That was the last autograph I got, not counting those from friends and acquaintances—well, not so, not quite. At a Pan American literary dinner where Pablo Neruda was the speaker and for openers said "*My Chee*-lay" and got a round of applause, I went up to the speaker's table with a program afterward and Arthur Miller, whose chair I had to squeeze past to get to Neruda, gave an appalled look, as if I were some sleazy, heartsick autograph hound.

"It isn't for me," I said to Neruda, in a tone all but accusatory after Miller's look, and Neruda frowned. But he signed. It was for a poet in the backwoods who felt Neruda was the closest you could get to God, and for the sake of us all I hope he still has it. A reception for Neruda was held that night at a South American Cultural Center, where I felt uneasy among the black ties. Then I saw Robert Bly, in jeans, with a serape over a work shirt, examining books on a table, my first glimpse of him in the flesh. I went over and said, "Is the North taking over?"

He turned with eyes so luminous their interiors seemed incandescent, looking high as a kite, and said, "I am."

My missteps in the vicinity of Parnassus kept up. After I married and found an apartment for us in Brooklyn Heights, the people who were moving out said Truman Capote lived on our street—down a ways in a yellow house redone by Oliver Smith, the set designer. Capote was finishing *In Cold Blood* and soon moved to the new U.N. Plaza Towers. But we met a decade later in North Dakota, of all places, where he was the headliner at a literary festival and I was one of the others. We both flew in from New York and the buzz was that he left his room only to do what his contract required and was demanding a baby-pink spot for his reading, at which he expected a hundred people from Kansas.

When I shook his hand I told him how I had read *In Cold Blood* in a horrified pant—its victims were the people of the safe countryside I knew—waiting for each section as the book appeared over the weeks in *The New Yorker.*

He said, "Some of your work is very sweet, Larry. You're not the guy I imagined. I like you."

That night at the hotel where the participants in the festival were housed (our contracts had it that we were *expected* to attend "social functions"), a raucous party started cranking up. "Get Capote here," somebody said to me. "He owes."

I picked up the phone and asked for his room. A radio was going, somebody had brought in a tape player that was turned up high, Ed McClanahan was whanging away on a guitar, and a couple in the corner looked like they were trying to smell a pocket mirror. A weak voice said "Yes?"

"It's Larry, Truman. A party's going on and everybody wants you to join us."

"I can't."

"I'll come and escort you and make sure you get away when you want."

"That's sweet, Larry, and if it was just you I would, but—" He seemed to assess the din at my back. "I just took a pile of pills, honey, and put on my eye mask, and I'm all but dead to the world, as I'll soon be. I have people coming to see me to-morrow. Tell them that. Bye."

At the conference Capote said that Mailer, a long-time neighbor, had stolen the idea of the non-fiction novel from him: "Norman for the life of him couldn't have *imagined* a book like *Armies of the Night* if he hadn't read *In Cold Blood*. But now that he's got the swing of it, boy! you'd think he had a corner on the market!"

The Kansans, only a carload of them, laughed.

Mailer lived at an angle from us on Columbia Heights, just off the promenade, and if my wife and I were on our stoop as he passed—usually to visit his parents, who lived in an apartment a few doors down—he said hello. We often had evening meals at the same restaurant, Armando's. He had a gentlemanly graciousness and sometimes paused for a few words of an everyday sort, but usually when we passed he merely gave a grunt of acknowledgment, a nod.

The year my first novel came out, he was running for mayor of New York and I encountered him at a subway entrance, where he was handing out leaflets to the crowd climbing up. "How's the book doing?" he asked.

"In its twelfth printing."

"Good for you!"

"They're small printings."

"Go get 'em," he said, grinning in a way that drew his eyes open wider, rolling his hairline back, and then he took my hand and shook it, bearing his vulnerability in a macho husk.

The storm lasts three days and nights and piles so much snow in our yard we can't get out with the pickup. The road is drifted over anyway, with bulges like surfacing whales across it, as we see when the snow lifts, so our biker woodcutter can't make a run. We're down to the last wood after our snowbound days, unable to drive even two miles to the cottonwood grove with its owl in the Quonset.

I lie on the floor in the greenhouse where we earlier installed the drain to empty the loops under my office, waiting for Joseph to call "OK!" This will mean he's turned on the spigots at the furnace for the lines to my building, and when he calls I switch on a pump. Only a dull buzz. We have to find where the water in the copper is frozen. The plumber suggested using antifreeze in the system but Filer said, "No, you have to drain the furnace every two years to clean it and the cost for enough antifreeze to do any good will run you a thousand each time. And no matter what they say, it's corrosive."

So we remove the floor in the entry, to expose the lines of copper there, remove the floor in a step-up level beside it, where I plan to install a washer and drier as backups to a pair in the basement; and finally, after these removals, I pass over the copper the flame from our map-gas torch, hotter than propane.

Beads of crystalline water, as if drawn through the copper, appear; lengths gurgle or jerk with the *wham!* of a steam pipe. Joseph crouches in the greenhouse, watching for water from our drain. I'm sure I hear the stutter of a boil at an elbow and back off from it (though elbows are likely culprits) for fear of loosening the solder. I'm ready to give up, certain the poly hose is frozen underground, when Joseph cries, "Dad, it's seeping! Now it's *running*! We got it!"

Neighbors Now New

Joseph wades west of the house with a chain saw in one hand into the scrubby triple row of trees on this side of a caragana hedge. He starts the saw by tossing it away from him as he holds its starter pull, then reaches up and trims dead limbs from a dying ash in the shelter belt—the only remains of an original planting on three sides of the house.

Most were Chinese elm, a sixty-year tree, short-lived, and we've downed and burned the dead ones—Joseph only the third generation, counting me, to cut trees on the place. When we moved in I spent a spring and summer trimming the tops of a hundred damaged by the tornado that tossed buildings around. It was that tornado, the Berns' second, and the wicked winter of 1976–77 (but not nearly as bad as this one) that sent the elderly brother and sister to town.

Five days before Christmas, the phone rings and I walk away—my least favorite task—as my wife picks up. She listens so long I'm ready to indicate with sign language to hang up if it's a telephone hustler, when she says, "We're so sorry, Kenny. Of course. We'll do anything we can to help."

I call back to offer my condolences, then say, "I'm willing to try to drive over."

"Why would you do that?"

"Talk, if you want." His way of dealing with the worst is to talk, I know, from the times he and Valeria stopped in when the worst for them wouldn't give an inch.

"Only a fool would risk a trip in this. I'm OK so far. It's just when I think how this came, right in the middle of the worst storm yet, like with—"

I wait, dutiful listener, for *Valeria*, whose death came in the midst of a blizzard, so we had to take the long way around, on pavement, to attend her funeral in Lemmon.

"Don't move unless you have to, I say. I called the county shop and asked them to send the maintainer out, which by law they have to with a death in the family. Call the county shop, too, and ask them to hit your yard on the way out. Say you're doing my chores. Your wife said you would, so thanks.

"It's a memorial service, cremation, and Kiyoko wants it later, so people from Japan can fly in if they want, and then others who will talk about Brad have to get there—besides, they know how we've been hit here with this weather. I talked to Brad every day, then Kiyoko. She was the one who called and was so up-beat it about broke my heart, a real Christian daughter. 'Cremate *me!*' I said. 'Or throw my bag of bones in the pasture.'"

He sighs and stirs in me my greatest fear: that my wife or son will die before I do—any of them before me.

"Somebody from Sony will fly over, she thinks. Brad was second to the vice-president—in Sony-America, I mean—the only guy who could tell them what was said in the calls from Tokyo, and when so many lost their jobs after Ideo took over and the shake-up started, Brad if anything moved up. No way, they said; our New York office can't do without him.

"You should have seen how Sony got behind him when they heard. No limit on the money, they said, for any drugs or hospital or treatment he wanted to try, no matter where, France or Poland, you name it. 'Take them up on it!' I said, '*I* would!' and he said, 'Dad, I think it's too late.' He didn't quite make it to Christmas—"

I hear him inhale moisture up his nose and wait. "You know the chores, the ones you've done, but this winter I'm worried about my pheasants. Keep those thirty gallon drums around the buildings filled with oats and keep the snow cleared from the bottom holes so the oats runs. It'll be a week."

He hangs up.

I turn to my wife, who stares at me, then past me, as if somebody has appeared at my back in the flesh.

I met John Updike, after years of sharing the same editor, at his 1968 reading at the 92nd St. Y. He chose a story from *Pigeon Feathers* that has a young man riding a bus and reviewing dates from history to match the city's numbered blocks and when the bus reaches the block that should represent the year of his birth, it's torn down. The audience whooped with delight at this.

Afterward a crowd was at his elbows for a cookie-punch reception, in the room where a bust of Dylan Thomas stands on a pedestal, and onlookers were lighting up. One could smoke in those days. I was out of matches and the pair I persuaded to attend with me, my wife and Bob De Niro, didn't smoke. My wife was in her seventh month of pregnancy and Bob so busy with work starting to come in I had to promise I would try to make a way for him to speak to Updike. He was interested in the movie rights to *Rabbit, Run*.

The crowd was breaking up for the real reception, at an apartment whose address people looked willing to kill to learn, and I stepped over to Updike, gripping a wrinkled cigarette I had been worrying since the end of his reading, held it up and said, "Do you have a match?"

Not a flicker from him of the pun, Superman jokes and the rest, which didn't occur to me until I said it, but instead his smile, overcrowded teeth thick as knobs. "I'm happy to meet you. Bill has mentioned you." Our editor, Maxwell. "No, I gave up smoking a month ago and of course now I think everybody should. You should. I haven't felt better in my life."

I quit for weeks and months, even a year, but an abrading concern, especially with this weather, is that I still smoke—a reason I run to town so much. I don't like to buy them by the carton, because that would look like I was going to continue to smoke. But I continue to smoke.

So I'm relieved when I see a road grader with a V-plow up front and adjustable wing plows (to knock down banks along the roadside) take three runs at the drifts choking the base of our lane and then come barreling up its incline. The driver blasts toward me, throttle wide, then lurches right with a squealing crunch. He's hit a buried stump near the circular drive, I realize. He backs, raises his blade to reveal shredded wood, then blasts straight over our yard (my wife!) to the garage-granary. Joseph uses a snowblower from the fifties to cut a path, shoulder deep now, for our car. The monster maintainer takes two more swipes—past the house, toward another garage to the north—and is done, the driver inside waving on his way out our lane toward Kenny's.

I drive over to meet my neighbor in a new way, and discover snow piled in his yard in twelve-foot mounds, the first I've seen it so, my first trip here since winter hit, and step into a frozen wasteland. No sound. Not even pheasants—the cock's rale of a rusty nail gargled. I knock on the door, wait; knock again. After Valeria died Kenny started sleeping on the downstairs couch, he told me, where she spent her last months. I try the door. It opens with a collapsing sound against a wedge of snow. "Kenny?"

The inner door to the entry opens on a silver-haired woman. I'm astonished at memory's resistance to change. In the summer, only months ago, with Brad's encouragement, Kenny married Dorothy, a widow at the head of a list Valeria had prepared for him of women he could marry.

"He's doing something outside."

"How's he taking this?"

"Oh." Her eyes work back and forth in her consideration of this. "Maybe he's not so bad right now. I'd like you to talk to him."

"I'll see."

A sound comes from a Quonset where Valeria once kept milk goats, an irregular banging only a human being would make, and I head toward the building. I fear, as you do in this country when the bottom is gone on the price of crops and cattle, and creditors are going lengths to get your last dollar—not to mention a siege of weather like this, besides the death in two winters of your two closest—I fear the suicidal swing, boots battering, from a rope strung over rafters; the bark of a deer rifle, shotgun thud, or a nap in the deep snow.

The sliding doors are parted but it's too dark to see inside the shed—morning glare in spite of no visible sun. "Oh," I hear, and walk in enough to make out Kenny, in a cage of wire panels, forking alfalfa to the center of the shed. "You'll need to pull this loose for the beef steer," he says. "I can't get the tractor out to feed him."

"I'm sorry, Kenny," I say, my breath pouring in a gray fan at him. Then in fear from what I've sensed, I say, "The Lord will sustain you. Dorothy will help."

"I know." He shoves the glittery tines of his fork into the hay but keeps hold of its handle—fuzzily flimsy orange gloves on his hands. "What gets me is how it's all changed, everything. It used to be if something like this happened, the neighbors'd be over, they'd get here somehow, and the family ahead of them. Not a person from the family has called! On my side or Valeria's! Hardly anybody has!"

The worst thing I can say, I sense, is I have.

He shakes with anger that colors his face but then releases the fork and sighs. "Did I tell you what Brad said to those experts and wise-asses in New York?"

"I'm not sure."

"They asked him how he knew everything he did, about electronics and TV and computers—he knew more than most guys with degrees in that. And he knew mechanics, so he could put in sound systems for these hotshots nobody else could get close to with a ten-foot pole. He put the system in the yacht George Bush used to piddle up and down the Potomac in when he was President—not that I ever cared for that mealy-mouthed double-crossing son-of-a—

"Excuse me. Brad got along so well with people he was the one they sent when there was a bad situation or delicate negotiation up the road, so somebody— People at Sony used to ask him, he told me, 'Brad, where did you learn all this stuff?' And you know what he said? 'I learned everything I know from my dad, on our farm in North Dakota!'"

His cry is so sudden I hear my father's wail of grief. But stand paralyzed, Kenny does—Germans drilled in the art of forsaking emotion. He puts a hand over his face as if to shove his noise back and I pray for release from the immobilizing freeze that all the deaths that began with my mother's have slammed into me, then go on legs like logs and put an arm around his waist. He shakes and wails but won't touch me.

This goes on until he taps my back like a referee, a signal to release him.

We get through that storm and then one pollinated by rain from a rise in temperature crosses the area. This is the week my wife drives to Lemmon for supplies, and when she starts backing out of the Jensens' yard, so filled with snow she can't turn around—trying to keep a cup of cocoa Sue Jensen has given her balanced in her lap—she slides down their approach into the ditch and snags a steel post that tears off a length of trim and sends a jagged gouge down her luxurious old car, causing me to say to her, once she's back, that we might as well drive it off a cliff. Because of the snow that's kept accumulating since, sifting under the garage door in a drift, the car remains in the garage, like a wounded lion in a cage, for two weeks.

Then the day of warmth when I make my excursion to the bank, with Joseph and Ruth off to work at Lemmon in the pickup. Stuck. Then back here I call John and a woman's voice comes on, his wife or a daughter, and when I ask if he's in and learn that he isn't, and wonder when he will be, the voice says, "Don't ask me."

I hope he isn't caught out cutting wood. The phone rings, Joseph: "People are saying we shouldn't drive back."

"I think they're right. I sure don't like the looks of this. Maybe you should stay with the Jensens."

And I'm back with both feet in the day this began, when it wasn't my hat or my head I forgot, bundled like a baby at the furnace. I dig down in a four-foot bank for buried chunks of cottonwood I hope I missed. A few. I pause in my stoking of the furnace, standing in baking desert heat, the open door a glowing window in the cascade of snow, where I see the future piling over our past and hear the icy chill in the wind I heard on the hill.

Internal and external weathers enter unbelief in a flying wedge: a sudden fault in your feet and legs, mine now trembling. You hear tinkling in a cupboard and the fixture above jerks, then sways, and you're in an earthquake. So when a wind like this picks up while snow piles in and then in a sudden shift everything is a squall of white, you're at the epicenter of a storm of earthquake proportions. It's here, it's above and around and under me, and I have no son to help.

The storm hasn't slackened by noon the next day so I phone Joseph and Ruth and I say they can't consider coming back. By now we're really out of wood, good cured wood, and the only worse element than uncured wood is no oxygen for burning. In my usual manner of denial or optimism, I think, *This has to let up*. I call John, who is OK, and says in his Mississippi accent, "It's cookin up real good here but if it gives in some I'll get out in it and get you some wood."

I walk our tree rows in what feels like a sandstorm, gritty burrowings even through my face mask, but all white, and find a half dead ash, V-ing into a pair of weak trunks. I cut down the one sprigged with stubby bare twigs and find I can't run the saw without the fingers of my right hand, my control on leverage and the throttle, going numb.

I run to the house and plunge both red hands in cold water. The unthawing, as folks say, starts with a sovereign ache like a glove, then turns to sewing-machine pinpoints of particularized pain where I got it the worst, in my fingertips. After a while of this, notching up the water heat, my fingers grow dully swollen but with enough sense of touch to manage the saw, so I go back out.

I cut the half-tree in lengths, that much. But then I have to carry each one a hundred yards to the furnace, through drifts so deep I sink past my knees with the weight of the wood and me. I discover that when I bend below the crusty surface the knee of the leg I used for my last step, I'm able to get enough purchase for the next leap ahead. Then sink in past my knee again. After an hour I have enough chunks at the furnace for the night. I load it and go to bed, leaving on my underskin of silk, warmed by *her,* so weary from working in the wind and freeze I'm out like that.

2

I wake and check the digital clock—3:00 A.M., black dark—and know something is wrong. Our bedroom is as cold as it's been all winter, including the day and night of no electricity. The wind has throttled up to such a gale the room feels insubstantial from drafts, a new sensation. I get on the rest of my clothes, pull on my boots and coveralls, my face mask and wool cap, and go out.

I know it's below zero when the hair in my nostrils is Brillo at my first breath. In cold like this you don't breathe through your mouth or your lungs feel scorched, as if you inhaled a wigwam of smoke. I sense the scorch though I haven't breathed that way.

A chill grips my back as I realize I'm at the epicenter of something worse. I can't see the closest building, not even the furnace, and think of the tales of the farmers who had their wives pay out a rope tied to their waists as they left for chores, so they were certain to find their way back, and some didn't. So their wives went to look and were lost, too.

Light will be my guide, I think, as I dimly make out a shifting glow from the bulb on the furnace. To establish this end of the rope I turn on the light above our front steps. The corner of the house is a lee, I find, as I step into the tearing snow. Instantly my forehead hurts. I put a glove over it and lower my head and seem to mark time in the wind coming straight in from the north, the approximate direction I want to go.

I get to the furnace, out of breath, out of air, about to lean against it when I see that the snow, winding in such a whirlpool it draws sheets straight up past the furnace light, has been plastering the air-tight door and melting, forming a nubbly coating of ice. Stalactites the size of my arm, stained yellow in an overflow from combustion, I tell myself, run from the bottom of the door to the ground. I grab the handle and jerk and the door pops open with a clatter of glass smashed.

The fire is almost out, the gauge down to 90. *How?*

On the side panel I flip the switch that turns the solenoid on and off. Nothing. I should hear the damper clang when I turn it off, because it should be open for burning, with the temperature of the furnace so low. Has the circuit breaker blown? Not with the light on the same circuit burning. I seem to remember reading about a smaller fuse inside the control box, and internally rail at myself for not ordering an extra.

But if it is out I can apply the Filer technique—the one we used with no electricity—take the grille off and prop open the damper door. Which means monitoring the furnace every hour, at least, to make sure a boil doesn't begin, and I'm so worn from working all day in the cold I'm not sure I can do that. Not in this weather. The wind is up to such a pitch it's all I hear. I start for the house, whisked behind in a way I don't want to be with the snow so deep, and bend over with a glove above my eyes like a scout, as if I'll be able to see what I'm about to hit that way.

Inside it feels like suburban Detroit in December. The batteries in most of our flashlights are dead, I remember, used up over the time of no electricity, and I forgot to get replacements on my bank trip. Two pocket flashlights work, I discover, the better of them the one my wife keeps in her purse, pathetic. But two lights as the ends of my rope so far have worked. I tread light through the kitchen, trying to keep from shedding snow ("It's clean snow!" people cry, when they hate its messy melt) and jog down to the basement. Somewhere is the right socket for the bolts that hold the louvered cover on, and now I at least have light.

Tools are scattered from our final installations and the frenzy of trying to keep everything from freezing with no electricity. The socket set is in the garage-granary, I realize, and find one of the mini-flashlights, take a final breath of normal air, and step outside. On my way to the garage I have to climb as high as the furnace over a drift, and once there, inside, I draw the door back down. Even here it feels warm. The car looks stalled in a ditch again, from the drift around it. Outside, socket in hand, I climb the new drift, which steps down from one that yesterday reached the eave of the garage, and the best I can do is crawl, then roll to the furnace.

To remove the louvered cover I have to take off a glove, and just as my hand is going numb I dimly see a difficulty in the glow of the furnace light and my mini beam. Through a silver-gray sheen that appears to flicker and dim as it thickens, a membranous layer of active snow, I make out ice at the base of the damper door.

I go to the house for the other pathetic flashlight, the last one, grab a huge screwdriver, and return to the curved cocoon carved by wind in the snowbank circling the furnace. The ice is so hard it takes fifteen minutes to chip it away with the screwdriver, but at least my hand has recovered from the trip to the house, and feeling kicks in at the work. I flip the switch and the door doesn't clang. I check the gauge on the furnace, down to 80. With this wind the pipes in the house will soon be affected, because in a kind of counter wrap of memory to the cold I'm caught in, I remember how cool the basement felt.

A mechanical part of the damper seems frozen, perhaps the rod that acts as a lifter to push it open, because I can see it bow with strain when I toggle the switch. I turn and head for the dim glow on the house and once inside I head down to the basement. Yes, cold. I find what I'm looking for—the map-gas soldering torch. I fire it up in the entry and the second I step out the door it's snuffed, the wind. I let it leak gas all the way to the furnace, jerk open the door again growing encrustations, and hold the torch near the last of the coals. It springs to life. I run the tip of its flame over the linkage, afraid of getting too near the top, which might burn up what I think is the solenoid, then remove the screwdriver, which I've been prying beneath the vent door with, and try the switch again.

Nothing.

I bang on the vent with the screwdriver handle, lave it with heat from the torch, and watch drips of water freeze before they're full enough to fall, that cold. I lie on the ground, out of the worst of it in my cocoon, and find I can reach the torch close enough to the damper from there. Then I watch as blowing snow starts to form a fuzzy layer over my coveralls, forming a drift of me.

I'm not sure how long I wait for the torch to do some good but soon I know I have to sleep. Life, brief as a breath, over? Carole, Newlyn, Joseph, Ruth, Laurel, it may be by a row of words you remember me, or maybe not. Or images, once my body is gone. You'll have to resolve the distinctions between those for yourselves, or you in Lemmon for us here, if I can't keep the torch on target, get us heat, undo the miscues that brought us to this and leave off! so you'll know it wasn't my interior and its revolving search for words that held me here, but you.

In Illinois our lives shift like the shifting sand of the dunes and sand hills in the one-horse, four-house town of Bishop our grandparents mean to move from. But once our sandlike settling begins, in the fall of 1950, it seems that the spot, an unattractive house at the center of Manito, draws my mother straight down. In three months she's dead.

I didn't know the details until this year, with the nun's clipping. All a child registers is the parent gone, *she* is, and then a wind that even for January feels bitter in Illinois is sending the sides of the canvas shelter we stand under flapping, its roof bowing up, then going flat with resounding concussions, rifle reports, as I try to listen to the priest and stare at a cumbersome drawer of polished wood, like a confessional tipped on its side to spill the rot of sin, that holds my mother.

And if it is this cold where we stand, I can't help thinking of her where she is and, worse, inside the ground, in the short-sleeved dress she was in at the overheated funeral home. She hates the cold. Drafts in the house raise such goose pimples, as she calls them, on her arms she has to rub them up and down to turn them smooth.

I can't get my breath, for her sake, and feel I'll gag at the thought that I might as well be lowered in the ground with her. The only days with any clarity are Sunday afternoons when our father takes Dan and me to the grave and we kneel as he does on the grass or leaves or dusting of snow, his head bowed, ours bowed, silent, a rosary wrapped around his hand and an ancient and staggering oak at his back if I look up, and when I do I see his pale face lifted, lips apart, as if life is pouring from him.

He reaches to her grave and brushes away leaves, pulls a weed, then another, smooths the sandy dirt where popping roots dislodge it, and on Easter lays yellow tulips at her head, on Memorial Day a wreath, and in the summer mows the grass at her edges and finally takes over seeing to the whole cemetery. Dan and I go with him, then Chuck, and our practical tending to the growth above her, in the swelter of Illinois heat, draws me back among the living again.

I lie in snow with a torch overhead and hear the damper snap open with a clang. I'm so startled I'm on my feet in a jump. It's worked. I slide the grille in place without attempting to fasten it down, and run in a stiff trot with the torch, hissing but out in the wind, to the house. I turn off its valve in the entry, pull off my coveralls, my boots, scarves, mask, gloves, jacket, socks, dropping them where they fall, and head for the warmth of my wife.

In bed, heated by her, I imagine wildlife in the cold—pheasants burrowing under snowbanks, where they smother in this weather, coyotes after them, so cold they don't yodel. *Our* animals. The sheep have hay enough but I have to check the horses in the morning—so much to do without children. Their favorite horse, because they grew up under him, literally, is patient Cody, a bay gelding, twenty-five and suffering the cold in his bum hip, where he was kicked by a stallion who broke out of a pasture, causing Cody to defend the mare he assumed was his. Alpha horse. Then the dog and the cats—four of them, three kittens from our female this fall.

Joseph cares less for animals than he did, and perhaps it's being nineteen, and a hunter. This year he got his deer just as a storm hit, and had to clean and quarter it alone in the garage, then used a hose to clean up, so that under all the snow, under the car, outside the overhead door, is a layer of blood that unsettles me.

Maybe him, too. He looked pale, apologetic, at the end.

Ruth cares more, though she loses her temper with a horse, especially if it pushes her, an aggression she won't tolerate. She was so energized, active and talkative it was difficult for us (I draw your heat closer) to keep up. Joseph had his stable of metal toys, inherited from cousins, and one, a bulldozer minus its treads and scoop, Ruth took as hers. She used it in the sandbox we built from railroad ties—under the shade of the Russian olive spread in such an aesthetic way our musician friend from Japan, Matsuda, said if he were us he would take a picture of it, with the children under it, every six months.

Ruth transported her bulldozer to the sidewalks—from the drive to the front steps, along that side of the house and around back, the branch from the steps to where I work—and ran it so wildly in a crouch, leaning her weight on it, the bare wheels sparking on the cement, I was afraid she'd fall. "Buh-doze! buh-doze!" she cried, when I asked her to slow down. But she couldn't, its speed exactly hers.

When she got on the swing set your dad and I, Care, set up, once our feeling toward each other started mellowing, she pumped herself as high as a swing went without wrapping or dropping her off upside down and would have tipped the metal frame if we hadn't embedded it in concrete. As it was she caused it to give and rock so much she loosened it.

Monkey bars ran across its top and from the time she was four she went hand over hand across these at her fast clip—"Dad! *Watch!*"—swinging around at the end and heading back without a catch in her pace. How strong she was, strong as a boy, hair flying, Scandinavian blond, a legacy from you! so thin at first I cradled her head against my chest with one hand, the other over it to shelter her from people who said "A boy?" or "Baldy." But, oh, like that singing swan when it came in! Fine and white-blond and platinum now.

I turn from you and slip into an earlier hour, when I was arranging the blankets you had draped over the doghouse (a project for the children years back) to hold in heat, and saw our female cat was gone. When I told you, you looked so dismayed I said, "She'll show up once the storm is over. She has before." And I said our unspayed female dog had anyway enrolled herself as the kittens' mother, as before.

All right, all right, I will before I sleep.

Out of bed I find the flashlight, pull on my coveralls, my boots, and out the back door shamble to the doghouse in a drift. Two of the three-quarter-size kittens are half asleep beside the dog, their faces and fur rimed white from the wind-whipped fog of snow no shelter can contain (lines are spreading from the doors of the house; drifts grow on the pickup seat), the dog's whiskers so laden they're drooping. My demi-beam catches the third kitten, on the dog's back, its paws curled under it in the way of a cat in the worst cold. All three blink back sleepily and the dog's eyes roll in shame, as though she knows no dog is supposed to enjoy a cat on its back.

I call Cody's name into the wind and imagine I hear him nicker, then I'm in bed, arranged along your heated contours with our wrap of thirty years, hardly able to steer myself in the direction of a prayer before the jabbering of those unfamiliar faces carries me toward sleep. And so I hang on again.

3

A radio is our alarm and we let it play most of the morning, hungry for news. This morning awful reports are arriving: a couple left for a destination so certain they called to say when they'd arrive but never did. Two people missing to the south, on Standing Rock Reservation. The storm has pushed cattle into fence corners and hundreds of head, maybe thousands (a helicopter engaged for the count had to turn back) have frozen to death where they stand. Last night it was forty below when I was out with the torch, I hear, with the wind climbing above forty miles an hour, setting the wind-chill in the minus eighties. Then I'm dead asleep.

I wake fatigued, and the idea of having to cut wood slides me into half sleep. Yesterday when I burned the half ash its butt ends sizzled in a sappy bleed, so I dug in every drift where wood was stacked earlier and found a few chunks to toss in and pep it up. I remember something else. I get up and dress, that ten-minute drill, and step outside, feeling my breathing shut down in the cold, worse today.

I push off. In the fall I felled an elm we watched die over the years, with not one leaf on it for I don't know how long, but its main trunk was so damp with sap I buzzed off only the limbs and burned them. I hope the main trunk has freeze-dried, as food does, and I go in leapfrog plunges—this foot, that—over drifts I stir enough for the wind to lift sandy snow and drive it through my ski mask. Past the caragana, I'm in a wind straight out of the west and before I'm halfway down the drive, in spite of all I wear, a headache of the kind ice cream can send in a shock from the roof of your mouth sets a silver wedge so deep above an eye I turn back.

Once warmed in the house, I hold a glove over the sinking silver wedge below my mask, climb the drift to the garage-granary, then climb through another inside to get to my work-bench and Plexiglas face shield for grinding steel. I have to remove my lamb's-wool cap to get on the headband of the shield and when I squeeze my cap over that it gives with a spurting rip. On my way past the car we'll have to shovel free from the building itself I see a curve of one of the buckets we set down, planning to continue plumbing here, when I said to Joseph, "Let's quit for now"—buried. We never got to that. The shield protects my face but I can't puff with the effort it takes to walk through the snow carrying the saw or it fogs.

A tipped end of the downed trunk swims above a drift, a rounded length of gray-black stippled white. I could dig to it but my hands already hurt, so with the saw going I cut down through snow to the tree till I feel the chain go into its free-wheeling spin, done. I move to the next length, my face shield streaking up, and finish as my fingers go numb.

I trot back, down to the basement, and hold my hands under running water at the laundry sink, moaning. Then go and dig out the chunks and feed the furnace and run to the house and peel off layers in the porch and fall into bed.

Oh, Ruth, Joseph, you work more than I knew!

I roll tighter in the comforter, wondering how we'll manage with them gone. Because they will go, as children do, and we'll have to. I see Joseph on his first assignment, trying to look over the dash of a pickup he drives to transport a drill to another field after I hooked him up, to save time during planting. The pickup moves in profile, his head held high but barely visible in the side window, opening an image in me—this is the way a story begins—of a tractor wheel in heavy rain seen through a basement window.

I left what seems the tropics of Illinois to go to my first year of high school in North Dakota and ended up in the fields of a farmer. When planting season went into full swing in April I moved in with his family and stayed through harvest. I was his hired hand but was paid to do what I most loved to do at that age, drive. The tractor was an Allis Chalmers WD–45, the medium-sized workhorse of its day, now a toy compared to the eight-wheelers drawing forty-foot toolbars, and my employer drove an Oliver 88.

I loved the Allis and rode its bucking polished-metal seat so many hours I got my fill of driving. During the busiest part of planting the farmer and I left the yard at sunrise and plowed past dark in the cold spring wind—no cabs on tractors then—by bobbing headlight beams. He was precise and demanding, a Bohemian, as he called himself, tall and powerful, with a "dem" and "dese" accent in tenor. I admired and feared his breakneck pace. Off the tractor he ran everywhere (barely forty, but at my age an ancient) and expected me to keep up.

He wasn't quite unyielding and had a sense of humor that went for the sexually explicit but wasn't too coarse for a fellow who wanted to learn. If a spring shower settled into a driving rain and the furrow my front tire followed started filling with water while the rear wheels took off in sprints of spinning in the gumbo that had the ability to turn as slick as grease, he pulled up beside me on his rig, running with his plows out in a high gear, and gave a thumbs-up: time to quit.

We parked in the yard, near the house, and ran to the basement. In one room was a regulation pool table he used to hone the skills he had learned in the army. The house had been built by a state legislator, three stories high and so spacious only two stories were used, and had the feel of an antebellum mansion from a history book. The eaves were fitted with solid gutters made of soldered copper flats (no round-bottom tin) with square copper downspouts leading to a central sluice that emptied the entire roof into a cistern.

This was poured concrete, one of its walls a wall of the room where the pool table stood, and the cascade in it grew so loud at points my employer would stop and stand with the butt of his cue stick on the floor, gripping it like a staff, and stare out a high window. It was one of the few times I saw him at rest, studying the rain, his face raised, attentive.

Outside the window a huge rubber-cleated tractor tire stood in the downpour, shedding streams of water over its ribs and hub, a wheel that carried us overland and allowed us to sow seeds that would, we hoped, grow into crops to harvest—none of this possible without rain. I stood in silence, too, watching him as he watched the storm the cistern beside us registered. Because he was so hell-bent on getting things done yet done according to Hoyle, I knew I was witness to a nature I would do well to copy: that listening angle of his face.

What I first began to attend to were the thoughts that overtook me on the tractor. Once I fell into the pattern of plowing my mind was free to range in any direction, a time farmers use to plan or mull over news and gossip. I had to attend to the direction of my thoughts in order to understand who I was and why I was made that way.

The similarities between farming and writing are obvious, when you think of how Tolstoy worked with serfs on his family estate at Yasnaya Polyana, or if you saw the dozens of varieties of roses Bill Maxwell raised on the grounds of his summer house, or heard a poet from the East told you he loved to weed his garden because it was like editing poetry. But it was in the fields of my employer and in his basement (where a cataract of water keeps clattering into the cistern) that I began to write.

I couldn't articulate it then but I was writing my way out of a former North Dakota, the seven-year span with my parents, into the space I would occupy as a young man.

The elm in the furnace bubbles, still juicy, and I wade to the garage-granary in the rising wind, a hand up to ward it off, and go to a pile of oak one-by-fours I've saved for a furniture project and sigh. I load the staves in my arms and head for the furnace through the wailing white.

In my last year of high school I stand over a man with the thinness of a diabetic, his skeletal jaw in a grip that sets grooves at his mouth, a streak of ink between a cheekbone and his ruffled hair. He is at a Linotype, in a steel chair that appears padded with rags, tilting a sheet of paper I've handed him—already fingerprinted by ink with its odor of alcohol—to better catch the light.

It is a poem, he is the editor of our local newspaper, and I need to see if my father is right. After he read it he turned to a shining black window, as if to compare my words to the night, the paper drifting down to one side in his hand, and said, "I think you've got something here."

The editor moves the sheet of paper under a gooseneck lamp glinting off the Linotype keys, holding it at arm's length as if the glow it takes on under the light is heat.

I'm breathless, as eager amateurs are, though they have the confidence of peacocks in displaying themselves, and finally he lets the page fall and turns to me.

"Hmm," he says, and his quick smile in the ink-stained aura looks neon. "Can you do this again?"

"Not like that, I don't think so, no."

"Why don't you try next week, and if it's close I'll give you a weekly column. Meantime I'll set this up for this week's issue, if that's OK, and if you have time with the rest you do—acting in school plays and all, I mean."

My pleasure isn't dimmed when I see how hard it is for him to fill the columns of his paper on his own. Over the year I write a poem a week and if my deadline is nearing with nothing done I pull down one of Dad's anthologies and head upstairs, break open a pack of cigarettes, and hit research in high gear. My poems and their tone follow the pattern of the form I choose and are windows onto what I'm reading: Poe, Coleridge, Lindsay, Frost, Penn Warren, Dickinson. It's a help to find Emily and learn that the best poetry is brief.

This is the main occupation of my last year in school and, besides the forms, I learn to write to meet a date. My weekly space is reserved and I have to fill it.

All cardboard boxes, all magazines, newspapers, all cartons holding historical issues of these, every scrap of hardwood less than yard long, a pair of two-by-twelves of rough-sawn oak I lugged all the way from Wisconsin—first I cut these in lengths—all this I burn to keep us alive.

In a college-English class I take under my father because he's the only one teaching it, we write an essay a week about what we read. We reach Shakespeare's plays, *Julius Caesar, Romeo and Juliet, Macbeth*, and the class says they can't understand this stuff, they don't get it, so he removes his suit jacket, loosens his tie, and starts to act and I sit back aghast that a grown man would throw himself into language like that. Once he moves into the play, he calls on others to read and soon we're inside it, no matter how some might stumble over words, and can see what's going on. We're there.

Near to the end of the year I write an essay that I begin by drawing lines that look like fractured Greek and after that describe what might be space aliens, repeat the drawn lines with slight additions, then describe the antics of the creatures; the lines further developed, and so on, recording events I experienced at the start of my freshman year in Sykeston, and after the final paragraph the lines form block letters that spell INITIATION.

"Did you find the idea for this somewhere?" my father asks. No, I say. "Well, where did you get it?"

Which is what surprised me, the way it arrived full-blown, in a flash, unlike the weekly poems I work at line by line to fill a form, although passages can fall in place in a cascade of inner recognition. In spite of the work on my column and my weekly essays, which my father usually likes, I'm not ready to admit that the emerging letters are like my emerging vocation, falling into place on the horizon.

Now that the surviving human beings have heat, I wade through wind to the shed we finished—screwing down the last panels of the roof through an ice sheet—and stand in its lee, able to breathe. I look in and see the sheep bunched in a far corner, ivory in the white fog of snow seep and their cottony breaths, the horses crowding them for warmth, their rumps facing the open door, not here to herd sheep. The sheep are a project for our children, registered Rambouillet who seem not to bother the horses, though I saw the pony Apache nail one in a swift mule kick, and with all that wool it went rolling like a puffball, jumped to its feet, and went *"Bah!"*

Snow blankets the horses' backs and icicles run from their nostrils and eyes. I brush them from Cody and he nudges me in gratitude but Dee, the head-shy buckskin, won't allow it. Apache turns aside in disgust. They have hay and water, the tank heater is working, sending up a cloud from the partly frozen water tank, and I've lugged along a bucket of oats I dump on a frozen bale. Then there it is, the reason they seem alien.

Their tails are hardly a foot long, frozen and snapped off in the cold, all equally smitten, awful.

How can they live with this, I start to think, and realize the convolutions of brain below my forehead must be freezing, too, because what I thought I meant to think was, *How can I live like this?*

INTERMISSION

Here in this noisy lobby like a gallery it's hard to talk, I mean above the clomp and gab, human thunder. On the wall behind you is the charcoal I mentioned or meant to of the poised woman with her chin in her hand, not as though she's thinking or at rest but at a ready. And there, farther down, the one you can't see as well, of a seated man with the top of his head sheared and preserved in turquoise. Catch the drape of his bulky tan cape.

I've seen some of it before, yes, or something similar, as I expect I will in the next act. But it's so hard to talk! You see, my life is spent listening, I think I said, and I listen to you, to others, usually, but replay it in private. My mouth forms words and a sentence starts to assemble. But I have to keep a kind of brake on the reflex that sends language after every thought.

Yes, if what I hope to get down doesn't take the wrap of words, what I do doesn't exist—at least not then. But the reflex can send up so many words out of habit they interfere with the process. Each sentence should have the freedom to assume its own shape. If I listen consciously or, worse, self-consciously, to a phrase that may be quaking in me with a sense of finality, it's gone.

Can we step over here? More has disappeared than I'll see in print but that didn't occur to me till it came out now. My source of work may be listening, but its outcome is loss. Once in print it's dead. And once you listen, assessing what's past, your source is silence. It's against a cyclorama of silence—you see I suck up what's close—that words start, and words are a way to find your route back. What was lost might be regained by them. The losses increase with age and after a siege of listening or onslaught like this I pray for the wits to sit to the end of one more act.

I know what you asked, and I think you should take this as preface. To hit anywhere close to an answer that's fair to us both, I'll put it in a story. When I was twelve and what happens to boys hadn't happened to me yet, I loved to walk. I would walk five miles down the railroad tracks to my grandparents' place or seven miles in the opposite direction to a lake I liked to look at, after I had walked to the far corners of our town a dozen times that day. It wasn't beyond me to walk twenty miles without stopping to think about it, as I haven't, really, till now, when your question returned me to it.

The places I most liked to walk were outside any habitation—in the gap of a railroad line or along a dirt track that led through pastures or rows of clattering corn to a woods. When I walked I thought of others who walked like this and the only ones I knew who walked as much as I seemed to were the apostles of Jesus Christ—along with Jesus himself, of course—and a U.S. President who once lived in the area of Illinois where our family carried on its transient state, Abraham Lincoln.

The place I liked to walk was a woods halfway between my grandparents and the lake, the straight north of those two points, although its actual direction, like most of my life, tended West. I walked toward it along the edge of a road of such rich sand it was as hard to negotiate as a beach. All along the route hedge apples lay like limes so bloated the pebbling of their peels had turned to worms locked in swollen combat. I didn't want to think what they were up to.

The hedge apples struck the sand like shot-puts and if I kicked one it left a gooey sap on my bare toes. Hedgerows crowded the road, growing wild in this place as deserted and hot as the Sahara—the perimeter of a state forest I was heading toward.

Once I had sized up my route for the next stretch, up to a hill or a curve, I never looked from my feet as I walked. I don't know why. What flowed past or flew in from the side or swung up to encounter me was more a surprise, I suspect. I partly wanted to be surprised, or scared, as boys that age do—a natural scare that would set aside the terror I lived with. My mother was dead and had died away from home, in a hospital, of a disease I had never been able to fathom or my father had never been able to explain or hadn't wanted to, so I had come to feel that my awful thoughts of her, years of them, caused her to die.

The latticework of shadow from the hedge-apple rows thickened to the shadows of trees, tall elms still free of Dutch elm blight, maples, burr oaks as gnarled as summer lightning, horse chestnut, and a dozen other types a teacher pointed out on a field trip when I was so taken by the trees themselves I couldn't retain their names.

But I knew them as well as aunts or uncles from my weekly walks through this forest that was also a wildlife preserve. I felt so much at home I sang as I did not sing anywhere else, notes that took the tone and pattern of plainsong—this I loved, mingled with incense, as much as anything about the church our father made sure we attended every week.

"Oh, beautiful trees!" I sang. "Oh, sky above! Oh, *earth* be*neath* my *feet*." It was a shout, blasts of a song I sang each time I walked here, as if announcing myself to the elements—the sky and the earth that had governed my life from the start.

And now these trees.

I was never afraid or lost my way no matter how many routes I took (besides not looking ahead) and never felt the absence of my mother as I did everywhere else. She was a girl of the plains who grew up far from actual woods—where an individual tree might set down shade or serve as a landmark but too many were a bother or threat. I had walked with her in the spaces of the plains and at the edges of the blue-green cascade of woods in Minnesota, and the angle and placement of her limbs as she walked were a revelation to me of her separateness. But people are made to talk, unlike the spaces of the earth (empty and filled-up) that seemed so much to want to speak they trembled with an omniscience that caused me to listen as I did not with people, not even her.

As I sang and walked, matching the words to my right-left stride, I saw rough trunks crowd close, their shadows lying on leaves or the needles they shed, their closeness intertwining in a way that caused the light striking my feet to take on substance. The chill of a presence slithered over me as if I were shedding leaves myself and I looked up. It was the presence of God, I thought, as I watched trees sway as if ascending the sky.

The patterns of the scribbled multitude of twigs and the matching gaps of designated light in sequence to the movement of the limbs were as much a song as mine. This was the earth, its trees in their multitude of beauty, twigs to branches to trunks, brimming with voices about to break into speech. I was in a grip greater than my mother's hand, and tears of laughter leaped out like the presences I expected to see.

One presence was here, I knew, as I turned with my face raised, in the trees and sky, and in the earth that held me as I turned. The presence had put all this in place to instruct me about myself and the complications of the love I felt for Him. I had been told to love Him but the words of the language I knew couldn't reproduce the language pouring from everything here with a familiarity I couldn't define. I was given a glimpse of it when I read, *The Heavens declare the glory of God . . . Day unto day utters speech . . . no speech or language where their voice is not heard . . . since the creation of the world His invisible attributes are clearly seen . . . All things were created through Him and for Him . . . and in Him all things consist.*

These were from languages partially put in English and what they stated in the English I knew imperfectly was so unimaginable—especially the last phrase—that I'm jolted out of that walk into this flat-footed noise. The last phrase is so seldom touched on by any segment of Christendom you would think it didn't exist, but the more I keep turning it every whichway it continues to say, *In Him all things consist*.

How do I reconcile that with my present-day pragmatic, tone-deaf view of nature—my hard-heartedness toward turf and trees and birds and fish and the stormy wind a psalm says fulfills His Word? Lopped off from the boy who hadn't learned to reason and didn't pay any more attention to his body and its developments (so far) than the developed creation around him, I seem further from the truth of the words of that statement than those who worshipped trees and imaginary spirits trapped in them.

And those who worship in this way miss the truth that the final statement and the ones before it partly catch: it is more than trees or the spirits in them if you receive them as attributes of a creator.

Carefully and with the greatest accuracy I may describe my favorite six-foot patch of ground or, if I'm feeling expansive, my favorite ten acres, and if anyone who reads what I end with doesn't feel in the description the hidden attributes we deny because they don't fit with the rationality that guards us from our fear of the unknown, then my description is a failure. Language was given to return to Him what He lays on those who only seem to hear, nowadays, when we're surrounded by a crowd or stereo equipment.

On some days, if I lie on the ground or writhe on my belly through grass or weeds or walk into a forest where I might get lost and lie down and take a nap and wake—on those days I sense voices clamoring from all sides as they did on my walks. How often do I roll in new-mown grass and feel its reek of greenness in my senses until it seems my nose will bleed and then realize the reek is the blood of grass, or something as astonishing, beyond my abilities to define?

There are times when, with a warning in my legs of spongy weakness, the earth is revealed as molecular—able to give way at any second, giving way—and every gesture and word and thought of mine is being measured on a set of balancing scales (right foot, left) to an accuracy of millimeters within infinity. The earth itself is embroidery and my treading on it is communicated through a network so complex even the mightiest computer can't estimate its effect.

This is faith.

In whom or what?

God.

Where is He?

Everywhere. Learned in English-language catechism.

Then why don't you bump into Him?

I do, in a sense, but on most days wouldn't know if He blocked my way, since I so seldom acknowledge His presence. If God is everywhere, it is as Spirit, and in this age of the Spirit in the world that came into being for that purpose, we don't accept the message His handiwork continues to communicate.

A glimpse of this was given to a poet in the Gulag. He was trying to read an anthology of modern English poetry with his faulty command of the language: "I remember sitting there in the small wooden shack, peering through the square, porthole-size window at the wet, muddy, dirt road with a few stray chickens on it, half believing what I'd just read, half wondering whether my grasp of English wasn't playing tricks on me. I had there a veritable boulder of an English-Russian dictionary, and I went through its pages time and again, checking every word, every allusion, hoping that they might spare me the meaning that stared at me from the page. I guess I was simply refusing to believe that way back in 1939 an English poet had said, 'Time . . . worships language,' and yet the world around was still what it was."

This is the Nobel laureate Joseph Brodsky trying to define the effect on him of a statement from Auden's "In Memory of W. B. Yeats"—*time worships language*. Brodsky hated sham and artifice and was so attuned to language, including the language of the Bible, that he saw Auden's statement as revolutionary enough to alter the world.

Time is the lesser compatriot to language and so time, whom most of us revere, bows to language. It is helpless before language. Language is greater, omniscient. With language people arrange adornments that endure, while time merely passes away. Brodsky grasped this on first sight under straitened conditions, but it flies past most of us with the thrumming beat of a flicker—the bird, I mean, with its yellow-gold ribs and flash of red you can't miss.

When I remember how I drew down great gulps Brodsky's explanation of his struggle to take in Auden's words, I see my feet moving through the woods and hear my song. And as Brodsky arrived by the labored steps of an unfamiliar language to an understanding of what he thought should alter the world, meanwhile transforming him, I had a similar sense when I wrote, "I feel a pressure behind and turn and there are the cottonwoods and willows at the far end of the street, along the edge of the lake, flying the maidenhair faces of their leaves into the wind, and beyond their crowns of trembling insubstantiality, across the lake dotted with cottonwood pollen, the blue and azure plain abuts against the horizon at infinity"—when I wrote that I knew I would never be the same. It was when the balancing scales beneath me were jiggling so much I was sure they would give way, and my search became a desire to rest, as if on a tree.

But I forget and become deadened, as I think I said, and walk around whispering, Sure, God's everywhere, that's why life's so wonderful—a detached and abstract cynicism so bitter it could burn holes in the air. When I got to a moment like that a year ago, my wife said, "Will you pray, please?" Sure, I thought, sure, I'll pray, and lit into a prayer with such anger a hole indeed seemed to burn in the air. Then it opened to the presence I had forgotten, and I felt the ladder of Jacob's vision, with angels ascending and descending on it, descend on me, the Spirit pouring though me with such force that prayers for my wife and children, who had gathered, were pressed from me as never in fifty years, and when I was spent and looked up I saw each of my family as if for the first time, transformed.

They were in Him, as I was, or more than I, *in whom we consist.* They had waited for this confirmation and I had been too cautious and rational and bitter (if I could have put my state in words) to give in to the Word Time knelt before. I had been reading Brother Lawrence, *The Practice of the Presence of God,* trying to develop gratitude for every breath of air and step I took, right foot, left, *thanks*, yes, *thanks*.

I went to bed. It was all I could do. I couldn't sleep. The pressure that caused me to turn and view the cottonwoods against the sky exerted such force I couldn't move. I lay beneath a molecular current containing yet bathing me, and every petty act of mine was like an electron above an abyss in a magnificence with no letup. I couldn't move for hours.

Every person I hated or could not forgive appeared over that time, not so I could see them, but I sensed the presence of each and knew who it was and was astonished at the smallness of my hate under the power passing through me. Tears sprang from my eyes as in the woods and I was lodged so close to joy I thought, If I have to die, if this is the end, so be it. And it was in a sense, perhaps (though you'll find me as unforgiving and petty as always, I'm sure, the next time we meet), because I understood I was being called to rise and walk.

I couldn't move but had to and once out of bed I made it from the room, barely (never waking my wife; she never woke the whole night), and slowly ground my way around the perimeter of two rooms and a hall, my pacing grounds, and had at least a partial sense of bearing in a body a weight like glory for those steps—all I could take. I knew I'd been prepared for this by that sense of pressure, that turn, but even more by the walks in the woods where I watched my feet lit by the sun as I listened to a language leaping past time and entering me in a way I couldn't begin to explain and never would have thought of trying to do until you gave me the opportunity, by your question, to reenter that walk.

II
Then with Tints of Snow

Onstage and Off

Summers are times of testing. I'm tested as I test what I've learned at school, now that I fall among farmers and reeves. I learn how universal laws apply to daily tasks and whether the words of poetry or prose fit the work. One summer in high school when I carry in my thigh the pin from an accident, then the scars and healing parted muscle after the pin is removed, an elderly farmer I run row-crop machinery for, says, "Good God! You walk like an old man! What'll you do when you're my age!"

Another summer I ride to a rural building site with an uncle—only the two of us at a framed-in house with stacks of bricks on pallets, a heap of sand, and a pile of mortar bags beneath a tarp. I mix and carry and he lays brick while he whistles a breathy catchall of songs, and at the end of the day, when all he has done are the corners, with a string stretched between, and one course laid along it, I think, *We'll be here a year!* But he keeps at it, his mild whistle going, and in three weeks we leave a house of solid brick—a lesson in the novel.

Sometimes in the midst of a reek of gummy wood on a framing crew, during lunch break, somebody will say, "What about one of those poems?" and I generally choose one of my variants on Mother Goose: "Hey diddle diddle, the cat and the fiddle, the cow jumped over the moon, the little dog laughed to see such a sport, and the cow got hung up on her bag." I have dozens, one for every rhyme, and want to collect them in a book called *Mother Goose-Um*. But I figure it would be better to do a serious book first, taken from my newspaper columns, published by my friend the editor, and one night I ask my father about this. "Let me think it over," he says.

I pester him until he finally says, "First"—a teacher's forefinger springs from his fist—"I don't think you should publish the book yourself. Whoever does it, if you pay for it, that's vanity publishing, and it never does you any good. Second"—the next finger—"I think you should wait till you're twenty-one to publish."

It isn't what I want to hear but in years will come to seem the best publishing advice I received.

The distances are what I like about the Urbana campus—elm-shaded walks of a mile to one building, a mile or more to the next, from Men's Old Gym past one end of the quad to the Vet Medicine Building past the other—a continuation of the walk I seem embarked upon. I also like having my brother Dan as roommate, in this community of 30,000. But it's clear I'm not academically inclined as he is. When I take my qualifying exam for freshman composition, the questions on grammar and spelling seem elementary, but we have to write an essay on the spot.

I set down as much as of the piece on initiation as I remember and it places me at bonehead level—not because I plagiarized myself, which they wouldn't know, but because it was, according to the note scrawled in its margin, "Not persuasive or argumentative, too creative. D+."

So to the bonehead class I go.

The summer after my freshman year at Urbana, I'm lying on my bed after carpentry, reading *Leaves of Grass*, which a friend suggested, when rhythms like Whitman's overtake me in a storm. I look up to locate their source and my head fills with an auditory echo of another rhythm that dims my vision. I'm sure it's my heart going, from the gray pall and noise in my ears, and can't move. But when I come to, recovered, the experience is so exhilarating I sit at the kitchen table with Dan late into the night and explain the meanings in a poem of mine until my jaw feels sore from talking so fast.

Then, a decade later, as I let my eyes run down a typed opening page of my second book, the auditory echo reappears, coupled with my voice, as if I were reading in an echo chamber, but the chamber throbs between my ears. And I realize what I've been led to from the moment with Whitman is my "voice." It turns on repeated words that I would normally revise from a sentence but that now seem raised ribs on the shafts of phrases that revolve within a sentence like camshafts, as they do when I speak.

My second year at Urbana I enroll under Rocco Fumento, who hikes himself up on the edge of his desk and talks in a soft voice fuzzed by his smile. Through a daze like naive innocence that clings to me and seems my Albatross, since it isn't actual innocence, I realize Fumento is the only person I would call jolly, his laugh is, as he talks about writing and writers. Some afternoons he scoots back on the desk, legs crossed at the ankles, and reads a story he thinks we need to hear from a magazine. He has published a novel.

A community greater than a newspaper editor exists here, though Fumento isn't part of it—so I gather from my hours in the K-room, across the street from Lincoln Hall and the English Building. The literary quarterly *Accent* was started by a writing teacher, Kerker Quinn, and the staff includes George Scouffas, Dan Curley, and Dr. Charles Shattuck, with offices in the second story of the English Building.

A sad-eyed grad who attended the Iowa Workshop and is back for his Ph.D. says Illinois lost its real writer, Bill Gass, when somebody in the administration complained about the food on his ties. *Accent* first published Gass and J. F. Powers and Flannery O'Connor and its staff is considered on the fringe, favoring writers like John Hawkes and Wallace Stevens and Eudora Welty. The editors often meet in a booth at the K-room, adding to the local literary heat.

The stately and dramatic Dr. Shattuck teaches Shaw and Shakespeare and directs plays on the main stage at Lincoln Hall. The rumor is that he quit college and ran off with a Shakespeare troupe that got in trouble with the law, then he came back and married an older woman who was married before and, besides his work at *Accent* and directing plays, has published three books. All these rumors, I learn, are true.

I enroll in his Shakespeare course, so popular his classroom is one of the largest, with seventy students. He walks in with the regal tread of a Shakespearean, wearing a trench coat and chocolate brown hat gripped so often the creases at its front cling, its crown sags. He stares down as if to locate a soliloquy as he sets a scuffed briefcase on his desk and pries open its jaws beside a wood pulpit on the desk. I sit in the back and take in the anticipatory quake of everybody suffering butterflies for him.

No need. Down with his hat on the desk, one hand in a glide over his silver hair, his trench coat dropped over the chair he moves aside from the desk, and out of the jaws of the briefcase comes the master to the G. B. Harrison *Collected Works* on the paddles of our desks, along with a file folder or sheaf of legal paper covered with his calligraphy—I checked it, under the pretense of asking a question, among the dozen at his desk after the bell.

He teaches as much by performance as his notes and the effect is of the class holding its breath. His aging face alters, the young actor in repertory appears, and he performs every important speech in every play we read, and then the scholar gives a pursed-lip explanation of what we have heard, along with quotes from scholars and, strangely (so it seems to me), his talk takes on a tinge of Shakespeare, a colloquial poetry akin to the poetry arriving in his actor's voice. It runs the gamut from high to low Shakespeare in one crack, in a unified voice, sometimes in one sentence, as "Let's turn to Act II, scene iv, and listen in on some of the modulated encomiums of Lear as I thrash hell out of it."

At times his hands tremble so terribly I wonder how he can turn a page or write on the board.

That year he holds tryouts for *Twelfth Night*, which he is staging on a reconstruction of the Globe. Years ago he had the set built from the specs of G. K. Chambers and others and it is used every other year—at least one Shakespeare on the main stage each season. One afternoon I'm walking through the scenery of my thoughts, dreaming of playing Feste, and feel a hand on my shoulder. Dr. Shattuck. He guides me across the street to the K-room, buys us both coffee and a roll, and sits me down in a booth. "Ah!" he says, then stares at the mug he's revolving. "Say you were Feste, what would you do?"

"Uh . . ." This is not done; faculty members who direct plays hold tryouts and post the results, and in the meantime ignore anxious actors. "Try to play him the way you mentioned."

"Good!" His teeth, splayed forward with the give that starts at fifty, the spaces between them wider at their tips, appear like a garland. "Good, good, good, but do you have your own ideas about Feste?"

I mention melancholy, which takes him to a campus visit of Thomas Mann. "It was when he was making his tour to let us know the Reich was as awful as we thought. There was a priest then at Newman Hall"—he jerks a thumb over his shoulder, toward the Catholic church across the alley—"a silly man who thought he was God's gift to university childer and perhaps homiletics, but a fool. He of course introduced Mann, who had folded himself into a chair too small for him, and our dear priest speechified about the honor bestowed in being able to introduce this great Mann, then helped us understand why, then complimented crumpled Mann on every book he wrote. Then he set into his real introduction, how terribly Mann had been treated by the Krauts—along with a host of others, of course, to let us know he knew what Pius XII was doing—and got so wound up you knew he was coming to his finale when, suddenly all red-faced, he cried, 'He's been crucified!'

"But that wasn't enough, quite, because he wound up again: 'He's been *crucified* on the *cross*!'—you could see him grasp for the thing to clinch this—'He's been crucified on the cross—*the cross of crucifixion!*' I was born a Catholic, you know, and if there was a hope I would return, there it ended."

I'm so afraid I won't get the part I can't laugh, though the lapsed Catholic in me coughs up some tries, and it must be my look that he wants, because he casts me as Feste. Most of the rest of the cast are regulars he uses, his troupe of grad-student insiders, as they're known, and I'm so intimidated I can barely speak when they watch. He eggs me on, demanding foolery, alertness, melancholy, and insists I can sing when I can't—and won't if others listen.

Rehearsals move from classrooms to the theater, the forestage of the Globe extending over its orchestra pit, so it seems I could touch the plaster decorations on the balconies above. He sits out from the stage at eye level, behind a plywood platform a janitor has rigged for him to fit over seat backs, establishing a table the size of a desktop, and employs a cigarette holder like a Hollywood mogul, waving aside tiers of smoke as he gives instructions for blocking or, screwing another Old Gold into it, sits with a contemplative fix on the stage, then moves to another sight-line, leaning back with a long leg hooked over a seat in front, at elegant Shakespearean ease.

2

The cottonwood grove didn't provide the best wood. It burned fast and hot and wasn't sufficiently cured—though the trees stood barkless for years, barren of leaves (that time of year that this in them you may behold)—so the remaining sap had to be cooked out before chunks burned well.

I searched the garage-granary for anything to burn. Junk disease. You catch it and it fits a farmer's failing, saving too much, until its weight pulls you into the pit of saving more. Which you do because you need so many parts and repairs for so many tasks each season you figure it all will eventually have a use, though it seldom does. Then you throw something away and need it the next week.

Cupboards and wood boxes and piles of parts and bolts, garden stakes and implements, a broken scoop with a new handle beside it—there wasn't a burnable thing that I didn't contemplate burning or measure against our survival—a fulminating greed to burn, but a greed discernment had to enter: I would not.

Yet as time went on, I must. Could I, I thought, first about the hardwood, then actual furniture I meant to mend, and then books, cartons of shelf-filling junk, awful books I knew we would never read, from farm auctions and library sales, books that the children of readers accumulate, never to read. All in a big box, into the furnace. Another one. Another. The Dickens I hoped to have rebound?

Something happens; I find a voice for Feste and Dr. Shattuck gives me a store of historical stage business for the role, but what I dread is when I speak a line or indulge in business so badly he strikes his forehead with the flat of his hand so the smack resounds through the playhouse. "No, no, no," he says, already in the aisle and heading down the steps, several at a time, in a catlike lope, then a hurried pounding up the stairs at the side of the forestage.

He first smiles his garland, to let me know he has my interest at heart, then takes hold of me and tries to get me to move or say a line closer to a conception he carries. I don't tell him I broke a leg so badly it had to be pinned and the point where the pin was removed still hurts so much I feel breakable, partly crippled, afraid he will wave this away like my singing or the pestering smoke.

Once when he demonstrates how a court fool like Feste would approach Olivia, he gives a skip that triggers him into a glide that ends in a kind of curtsey at Olivia's feet. I stand stunned. A magical fluidity has sent him flying, dissolving his years, and I sense I've seen Armin or another of the clowns he mentions in his class. I practice the maneuver in private and try to see myself uncoil with his elasticity, and at the next rehearsal in a skip and leap I'm at Olivia's side, down on a knee.

Smack, I hear. Then, "Where did you get that?"

"I was trying to do what you—"

"That's it! That's it!"

I become somebody else—Charlie Shattuck at nineteen, never so exhilarated in a role. *Otherworldly*. He cheers me on and the more he does the more my leaps and arabesques achieve the fluidity I saw in him, and I become so happy at getting things right I'm buoyant, airborne beyond the limits of a body. It isn't only my happiness, it's his; some nights he whoops and demonstrates another possibility, and the part expands. At the first dress rehearsal, as he arranges the curtain call, I keep expecting him to say, "You now, Feste," until after Malvolio's pompous strut I'm left last. "Now roar out dead center!" he cries, "and fly all the way up front!"

Which I do, which is the remarkable part.

On opening night, with the audience reacting as he does, I'm buoyed higher with each entrance, up to the last night, when the stage is bare and it's my turn for curtain call, I come belting from the inner stage, rising with a kick of my heels, and fly so far forward I hit the railing girding the forestage. The *Oooo!* of the audience is the kind you hear at a collision.

What I think I did was pull a piece of business, as if the impact were intentional, for another laugh. All I remember is a sound like a board struck by a sledge and how I'm curiously unhurt, and next being in the wings beside Dr. Shattuck, who is already all the way down from the top of the house, his hand on me. "Are you all right?" he asks, and I can feel the terrible trembling of his hand go through me like a chill.

"Ay," I say, and in that quickness I'm Feste, drawing us into one of those moments when a mystery is suddenly disclosed and as quickly buried. We start laughing as if it's the gliding hop he demonstrated that released me into this state, where I can leave myself behind and be somebody wholly other.

I write a story for Fumento's class overnight, when the image of a girl rises and spills out her situation. Then I sit on the floor, my typewriter between my legs, and type up the handwritten draft. The words are hers, with the weight of her way of seeing, and fall in place in a gradient and tone that isn't mine. I retype it and run to class to hand it in on time.

At our next meeting, as we're filing out, Fumento says, "Would you stay a moment, Mr. Woiwode?"

From a folder on his desk he pulls out some pages and says, "Did you write this?"

It's done on my typewriter, from the way particular letters bend, and then I see it's my story. "What do you mean?"

His jolly laugh sets his head back. "You have to understand I deal with all kinds. Last year I saw a story twice. Some fraternities have files. This is yours?"

"Yes."

"Did somebody give you help?"

"*No*." I'm offended.

"Odd," he says. "It's not the type of story I like—it depends so much on language—but it's the best I've seen in a while. You ought to enter the contest."

"Contest?"

"The English Department's writing contest. In short story it's the Leah Trelease Award. Money! Don't writers always need money?"

I show the story to others who say they don't get it or think it's crap. And when Fumento passes it out for workshop discussion it seems he spends his time trying to persuade the class it works. I finish my required three stories for the class, each more realistic and easier to understand, as Fumento likes, but the last one takes off near the end. My mother's voice.

Posters for the contest appear and I read the rules, no names on the MSS but a name on the envelope you submit them in, and so forth, and only so many pages. I retype the three, fixing them here and there, then count up the pages, and hand in the first and last.

With the close of *Twelfth Night* I feel the way actors do after a full role—afflicted with meaninglessness, a sick charade in search of a character. I meet a young woman with flossy blond hair on the stage crew who hopes to act, and I act as if I hope to import a definition of myself from her. We talk on every set of stone or marble steps on campus, even on inner stairs, having to move aside so others can pass, and as the weather warms we lie on the grass of the quad. She takes to chewing stems between her plump lips as we talk. She is a musician, but wants to be an actress or writer. She believes I can act and hopes I can help her, though I have no more idea of how I did what I did than a stone.

She practices Feste's leaps in the grass and enters the writing contest. She composes songs and plays them for me, the non-singer, on any piano available, in practice rooms at the music building, an old upright in a hallway, the grand at her sorority. One of the songs is based on a novel from her English class, *Zorba the Greek*, with such percussive thunder under its tinging melody it could be the theme for a movie, and I tell her so. She says she would rather write a musical, and the song she's playing, she decides, will be the opener.

"One about Zorba?"

"More like *The Music Man*."

"Seventy-six trombones!"

"That's it."

I have snippets of dialogue of the like in my final story, trying to see her side of how it is to try to communicate with a person in search of a character.

I pause over the carton of Dickens I intended to have rebound, recalling my instructor in Victorian Literature, Professor Smalley, editor of the Rinehart anthology we used, his name an oxymoron. He weighed three hundred pounds and his chief talent, besides his scholarship, was reading like Shattuck swatches of Arnold and Browning and, best of all, Dickens, when he became so engrossed in a character done in the suety sweet prose of Dickens, his whole body shook with amusement like St. Nicholas in labor. But no memories, even the best, will warm my family, so into the furnace.

Chickens are dropping in the cold, frozen stiff inside the hen house. I throw their carcasses in. Later, along with the stifling smell of burning feathers, an aroma of cooking chicken in the cold increases my hunger. I'm always hungry, as I was in college, or in New York when at times I felt I was living Hamsun's *Hunger*.

Dr. Shattuck walks up with a grin designed to conceal uneasiness and escorts me to the K-Room, then to a booth, and brings us coffee. "That's a lovely young woman I've seen you with," he says, and with the trace of agitation in him, this seems ominous. "Dear young man, I never thought your range of sensibility—I mean, I never taught creative writing with any success, though I think I developed a reasonable knack with my fellow editors on *Accent* for sensing what's good and what ain't. So often I'm asked to chair the departmental committee that awards the writing prizes. This year it was no question with the committee on short stories—the pair of stories submitted by 'number 19' were beyond measure the best of the season.

"Once we agreed on that, we hurried to the office to discover who the author of these brilliant revelations of a feminine imagination could be—so sensitive, so delicately phrased, so poetic, yes, in the highest sense!—I run out of terms we were using to characterize my nineteen-year-old feminine author, and in our hurry to find out who, this honestly came to me: that lovely young woman my feature actor has been captivated by!

"But no, Larry, it's *you!* You're the one. Dear Larry, good Larry, *fine* Larry, you've won it! How could you keep this from me? It finally dawned on me that the style, the mind and sensitivity behind it was yours, above all, but it reminded me of the novels of my dear compatriot on this campus, Billie Maxwell—especially *They Came Like Swallows*. Have you read it?"

"No."

"Come with me." He stands. "I'm adding a prize to the cash you've got coming." He turns, then puts a hand to my chest, his hat in the other. "Do not leak one word of this until the official announcement, not even to your lovely gal. Is *that* why I was so sure it was her, you're that close? All right, you may tell her, but not one soul other, and make her swear upon your sword to keep mum." A trembling finger goes to his lips.

He buys *They Came Like Swallows* and *The Folded Leaf* in a bookstore, in paperback Vintage copies, then out in the sun pulls a fountain pen from his suit jacket and writes my name in each, then, "Leah Trelease Short Story Prize Winner, 1961," and signs his name. He slides them back inside the bookstore bag and hands it to me. "There, and godspeed," he says, and goes striding away as if late for class.

I start for my dorm, off balance, dizzy, afloat among trees loping by on the sidewalk. I feel I should visit my inamorata and maybe muse (though I wrote the first story before we met), and I turn her way, but can't keep from opening the bag and taking out the book on top, *They Came Like Swallows*, and then its first sentence, a boy named Bunny, and suddenly I stumble on a curb, see that I'm in downtown Urbana, way off course, a mile from campus and farther from her—miles to go before I'm home.

. . .

On the phone I learn from Joseph that Highway 49, which runs north from Lemmon near our place, has been plowed, or so he's heard. The wind is up but not as bad so I say they should try to make it back. I blow out the tractor with the snowblower attached and go at the worst parts of the drive from the mailbox to the house. The yard itself is so packed there's no use doing anything with it. It's still so cold, with the wind, that my fingers are numb in minutes and I have to run in to warm up. Finally I finish enough so they can approach the house when they arrive.

Then we wait. The wind climbs, it falls, until it's an hour past the time they should be back, and then more, and I castigate myself for telling them to return, putting them in danger, merely because I can't keep up. I cut down posts, creosote-soaked, from a wood fence we disassembled in the fall, and pile them in the furnace. I dig down to the rough lumber that faced the fence, which I intended to use on another, pry them loose from ice, cut them, and toss them in. Then run inside to recover, and still they're not home.

And now it's dark, that quick. I go out and turn on the yard light and finally, faintly down the road, I see a corona of yellow take the bobbing rock of headlights. They continue in our direction until I want to yell "Praise God!" They pull into the yard and I run to the pickup, spent from the psychic effort of trying to hold close a child (no matter what age) when the child or children are gone, and find them acting grave, solemn, staring at the ground. No hugs. Ruth barely greets me.

"The banks the plow left on 49 were higher than the pickup," Joseph says. "And the snow was starting to drift back across. I didn't think we'd make it. I could hardly see the road half the time. It's been three hours."

Inside, I take Ruth in my arms, then Joseph, and ask forgiveness for insisting they return. "No," Joseph says, "we had to." So again I take hold of Ruth, who is called, so I hear in the family, "Daddy's girl," and smell ice, as if we've passed our time of tenderness and stormy days are here. I go into the bedroom and fall on the bed, finally able to sleep, but can't. I've tried to have a dimensional relationship with all three, no?

I once felt Ruth wasn't receiving the attention she should, because Joseph, only son, was three when she was born, drawing me his way when she needed a father's hand. Then at two Laurel arrived. Ruth has Scandinavian blue eyes like her mother, almond-shaped, nearly perfectly matched, set level at the start of her nose (which runs in a narrow line and then, at its tip, broadens), usually glittering with interest wherever she looks. All our children's eyes have a balanced, level, alert look, while mine with their flap of Mongol flesh at the corners draw down as if to stare at my nose. Ruth, at one year, called her source of food "norsers"—already inclining toward horses. At two she told a visitor, "Cody poky, Dee 'pooky," sizing up our herd.

My wife and I used to take rests together, the only time we had alone with three close children striving against the limitations of days and their ages and the urge to grow. One afternoon when I was away, my wife in the house, Ruth ran in crying, saying she fell from the monkey bars and hurt her arm. She was five and my wife thought she'd fallen on the branch protruding from her arm. It was her bone. Once it was reset surgically and put in a cast my concern for her turned to empathy because I once wore a cast and partly understood how she felt, and it was then I started "our talks," as I called them, holding her on a knee and speaking close to her ear in her rapid rush, to be one with her, and learned by this how fast she actually talked.

She was our historian, we said; if anybody asked a question she answered in entire honesty, with none of the social circumspection or scruples that crimp an adult sense of truth, and continued down other avenues until she was giving the details of all that happened in our household for weeks, including disputes and embarrassments. She seemed to preserve moments with a photographic integrity or better; laced through were networks of language from the stories she composed when she must have been busier than we knew.

She had a flawless sense of direction, so when she was four and I got lost in an urban area of New York or nearer home, she said, "No, Dad, turn here." Once she had been somewhere once she knew how to find her way back—her internal compass set at true north. She walked under horses from the time she could walk, setting my hair on end, even taking hold of a leg and swinging herself around it, with the sense of humor of the fearless, and of course the worst reaction around a horse is to hit the ceiling with fright, so I would say in a whisper with shaking under it, "Ruth? Ruth?"

We raised Leghorns until we learned the roosters were vicious and tended to pick on her—because of her hair, I suspected, the white shine of it. I kept one of the worst locked in the barn but once when I opened the door with her at my side it came flying from the inner blackness, claws first, and caught her with its spur below an eye. I always thought if it weren't swift Ruth the rooster would have had the eye it intended.

She had a gift with horses, an inner sense of their nature ("I'm part horse," she said) and got their attention right off and held it through the moves she imprinted. So we sent her to trainers who also saw the gift and this winter she has been riding every day when it's not thirty below, bundled up, her light complexion crimson in the cold. In her Scandinavian friendliness and interest in others, there isn't a hint of the xenophobia of our Germans and other locals. Xenophobia is too mild a word—not so much a fear as lack of charity for those you can't classify.

3

John Nims, a poet who reads and translates a half dozen languages, arrives in Urbana the next year, and I enroll in his poetry workshop. He is careful, unassuming, scholarly, a fastidious poet who totes stacks of books to class and reads from them, often repeating one line a dozen times, shifting its sound or sense with his emphases, teaching us to be as testingly tentative, his cherry-red lower lip rolled out from his swept-back chin, plumply tapping the upper.

His workshop meets once a week for three hours and he takes a break halfway through, so we can smoke, writers, you know, and sometimes he stands in the hall and puffs on a cigarette, gripping it from below with his forefinger and thumb, as a foreigner might, letting blue uninhaled smoke curl out cratery nostrils. During a break I ask if it's true that he translates seven languages: "Oh, some," he says, setting his lips. "A little, I guess"—unaffectedly modest.

In his workshop an older student with a brush mustache and mop of bran-brown hair, who is so thin his face is taut on its bones—wasted, to use a word he does—seems my nemesis. He picks my poems apart. These are typed up by a department secretary one after another on pages with no names attached and then are mimeographed, so that submissions arrive anonymous, in order to encourage unfettered talk.

Once when we're discussing a poem about a fledgling that's fallen from its nest, which is somewhat slight, I feel, and too loose ("How can you write free verse if you don't know what you're being free from?" Nims often asks, along with a Pound dictum, "Poetry should be at least as well written as prose"), the thin fellow says, "Yeah, well, that's my poem you're talking about"—a total breach of workshop etiquette.

The room is long and narrow, with conference tables in its center set end to end and barely enough space behind chair backs for a person to walk, but that's what the wasted fellow in wire-rim glasses, his face pale yellow after his pronouncement, does—hurries with shoves of grating chairs out the door. I'm afraid for him, partly from guilt at my criticism, though I didn't know the poem was his, and the walkout brings such a stunned wash of silence I decide to add to it. I get up and follow.

And see him crash through a rest-room door down the hall; I stand outside a second, then go in. He's bent over a urinal, throwing up. "Jeez," I say, "are you OK?"

He holds up a forefinger, goes into another spasm, then jerks brown paper towels from a dispenser and uses a wad of them to wipe his face.

"Did it get to you that much?"

"What?" he asks.

"The criticism."

"I think it's the fucking beer."

His name has an ominous overtone, Paul, but his friends call him Hap. Hap, he says, because he's always happy, and goes into a seizure of laughter, his head bobbing as he wheezes, his brush mustache of Theodore Roosevelt drawing back from bulged front teeth with a central gap, eyes watery. He grew up a hustler in the pool halls of Arlington, Virginia, his home, and is a caricaturist talented enough to make a living at it. He can whistle Shostakovich's Fifth and sing harmony to any song he hears. *Paul Tyner.*

He is working on a Master's in mathematics, drawn to topology and the four-color problem, the map-maker's dream, and no matter what hour of day or where he is, as in a dark bar in the afternoon, if you ask him the time he can tell it within a minute. "My inner clock is seldom wrong," he says. Then in imitation of Nims' plump lips testing the scansion of a line, "My *in*ner *clock* is *sel*dom wrong." Then he goes bobbing into his wheezing laugh.

Writers and actors congregate in the K-room, but Paul holes up in a booth at Stan's, a campus bar run by a former football player. Other writers start to gather there, most of them enrolled in an advanced story-writing course taught by George Scouffas. I'm taking it. A close-knit group of Lithuanians from Chicago, children of the generation that escaped Stalin's purges, is affiliated with Paul, through a young woman he dates, a blonde who wears her country's traditional embroidered jacket. Two Lithuanian chums in our class often use the word "existential." And "absurd."

One wrote a story about a fellow from Chicago who lives a pointless existence and in the last paragraph walks through a city park saying, "As I was strolling through the park one day, in the merry merry month of May," and then, without a hitch or transition, the next sentence and the last, goes, "How would you like a kick in the nuts?" Absurd. The story has a swaggering canter that Scouffas says is difficult to sustain, and though the ending might seem to some intended to shock, he finally felt it wasn't gratuitous. Three words Scouffas often uses are "gratuitous," "pedantic," and "grotesque," and in combination they seem to describe the intradepartmental politics he faces; he chose not to get a Ph.D. so the department has made it clear he won't advance, even though they know he's the finest writing instructor they have.

If Dr. Shattuck is dramaturgy personified, Scouffas is the pensive intellectual. When I smoke a pipe I think of him, imagining my chin as his—elegantly square, cleft down its center. Perhaps due to his pipe he has a shapely overbite, U-shaped, that adds to the impression he gives of tasting words before he speaks. He sees fiction with the eye of a critic, a fair one, able to judge its inner workings, as certain editors with a bent for formal distinctions can—split this chapter in half, cut those pages—while other editors enter in a writerly way the integrity (or lack of it) of the story as it stands and work from within, sensitive to false notes.

His techniques are simple. Until our first stories are in he reads from back issues of *Accent*, William Gass or Frank Holwerda, masters of the grotesque, then guides us into a seemly discussion of their work. When your story is ready you have to read it aloud, so you hear your words as the others do, their antennae out, and then receive what they have to say, as Scouffas moderates and adds his remarks. Then he edits your story as if it's entering the heaven of publication, and schedules a private conference with you.

The workshops are on the second floor of the Women's Old Gym, around an oak conference table used by the editors of *Accent*—one wall a jumble of books for review. Across the way French windows lead to a flat roof, resting on columns over a porch below, and a scent of chlorine from the basement pool adds an otherworldly addling to our tiredness. At this level of writing, everybody looks tired; the class can be grueling, though in control, as Scouffas uses the stem of his pipe like a pointer or warning tool.

The next story, Paul's, is about a pool hustler, Herby, turned cop in Washington, D.C. As Herby is strolling through the park one day, out of uniform, he recalls an immigrant calling him "shutzman," the one who does the shooting. Herby is on his way to a poolroom to win money before going on his beat. There we meet Sal, a slick shooter, and the bitter loser Buddy, along with a galley of hustlers and sharks. Herby loses to Sal and then, on his beat, sees a black man snatch a purse from an elderly woman in the park and takes off in pursuit. He catches up with the robber in an alley and as he holds his revolver on the man, Mark, he makes him answer questions—"What's a mark, Mark?"—and then shoots him in the face. This story Scouffas finds superb, for the authority behind it, the poolroom ambiance and the rest, then says, "I find it illustrative of the hate gnawing at America." It is 1962 and everybody in the story is white except Mark. Everybody in our workshop is white.

Three or four older writers go to the K-room afterward, and Paul invites me. He's an adept at nicknames and it may be then that he starts calling me "L." It's then, anyway, with his invitation to join the others, that I feel part of a community. The talk again is of absurdity but with a flavor of something else. I know, as the others do, that Paul's story is a kind we've never heard. That good.

I enjoy the private conferences with Scouffas, when he makes points, good and bad, he wouldn't in public, for your sake. The green-eyed dragon of literary jealousy has poisonous fangs, and he recognizes the precariousness of laying swatches of yourself on the table, because it's what he does, his almond-brown eyes taking you in so wholly he's vulnerable, your equal. He turns his chair to you, an elbow on his orderly desk, neglected pipe going out, and runs a hand though his thick dark hair—swept back like Cocteau's, with an irregular widow's peak just off center.

His look assumes you are responsible for your fictional world—the typed pages he grips at the edges with both hands, peeling back a page to a passage he's marked. His gift of encouragement is so subtle I don't realize how heartening it is until I see, close up, a shimmer begin in his eyes that means he will smile: that shapely overbite.

I encourage Paul to enter the writing contest but he doesn't seem interested and is seldom organized about his next strike—a Christian Scientist who can recite pages of Mary Baker Eddy and then knock back a beer and say in an old-woman's voice, "How the material asserts its devil blues in me!" Some days beer seems to poison him, until he heaves in the glass he's drinking from; then he braces himself and drinks it down. Whims sweep him and sometimes he sets down strings of titles, pages of them, as if he has a story for each. His favorite title is "Michael's Finger."

His next entry, which he taps out in his booth at Stan's, "Vittorio and the Llama," has only those characters, an elderly man who visits the Washington Zoo and, in his loneliness, talks to a llama. I feel it's better than the other—bittersweet, elegiac, stripped of irony, a gem. My own stories are becoming, well, more absurd. In one a pair of boots walks across a landscape scrupulously described (but new to me), then lines up with matching boots while the camera pans up, as it were, to the face of Abraham Lincoln. He is reviewing troops before a fatal battle. In another a woman visits a farm and talks about Jesus' second coming, never allowing anybody to get a word in edgewise, an elephantine unsightly farmer's wife in "split shoes"—a detail a theologian in the workshop calls obscene. When our group gathers in the K-room afterward Paul cries, "*Split shoes!* you dirty-minded little devil, you! I love it!"

. . .

Then Paul is in my rented room in a private house where a few weeks earlier the rangy actress from East Germany gave me the goods on my name. He is typing up a story he's obliged to stretch to meet a requirement for Scouffas.

Paul lives in Urbana with an uncle and aunt who won't let him work all night or have a beer, as now—Christian Scientists. Stan's is too raucous for him on the weekend but a buddy from there is present to help with dialogue. It's fairly clear the buddy and I won't hit it off when I pull a pack of Tareytons from my shirt pocket ("I'd rather fight than switch" is Tareyton's slogan, featuring print ads of men with black eyes) and he yells, "Pussy cigarettes! You smoke *pussy* cigarettes?" These are any with a filter, I learn, putting a finger to my lips for quiet. He smokes short Camels, plain.

The novella, as Paul calls it, is titled "A Bouquet of Helicopters." So that nobody mistakes his meaning he has pasted maple seed pods, the kind that whirligig down like copter blades, on its title page. When a character gets sunburned he pastes a sheet of skin he pulls from a sunburn of his own across the page. The novella is a series of disjointed scenes—episodic, Paul calls it, a Scouffas word—and his buddy from Stan's is more grotesquely wild than absurd. He invents a character called Marge, with a "uniboob" she throws over a shoulder, and they laugh at how to employ her anatomy.

"How about introducing a Mack truck," I say.

Paul turns to me, pale and unsmiling. Either in confidence, I now see, or when he was too loose for prudence, he told me he couldn't remember his father, who left home, Paul said, when he was four; his mother, a government employee, raised him. "But I have this recurring dream. A white Mack truck is coming toward me ninety miles an hour and I know it's *him*!"

I walk to the library or the Spudnut Shop to work and one evening I come back and open the door on a silent room, dim, its single shade pulled against an orangy sun. On my desk is *The Chateau* by William Maxwell, recently published, my one hardback of the year, which, again, has tripped me on a curb before I got home. But this book I'm savoring.

Then I see, stretched out on my only piece of furniture, a green Victorian couch with a carved coronet of hardwood across its top, a woman with her face to the couch's back, the curve of her hip rising high in cutoffs, her shapely legs drawn up to fit herself to the couch's length. I tiptoe over. Curly white-blond hair, rose complexion—the young woman who's been running with Paul's buddy from Stan's, the Camel man.

"Gone for a beer," a note on the typewriter reads, and on the page rising from its roller I see that Marge has been added—an affront to this painterly arrangement on my couch of the comely curves of womanhood, blond on green.

I pick up *The Chateau* and head off into the night.

The Main Reason to Breathe

At the end of that school term, I can return home and do carpentry, go to Breadloaf where John Nims teaches and has fixed me up with a scholarship, or join a Shakespeare repertory theater in Miami, at the invitation of the director, on the recommendation of Dr. Shattuck. I choose Miami, for one reason only: a play in their repertory, *Hamlet*.

But I get Guildenstern. Or is it Rosencrantz? Jay Robinson, who played Caligula in the Hollywood religious epic *The Robe* and became involved in a dissolute emperor's life in L.A., with a zoo of African animals outside his house and another zoo inside, plus an apothecary, is back in his hometown of Miami recovering. He is Hamlet. In *Twelfth Night* I play not Feste but

Toby Belch, maker of noise; in *Julius Caesar* Casca, a role that enables me to vent my discouragement about Hamlet, and in *The Tempest* Caliban. I work like Setebos on Caliban's slimy charm and toward the end of the summer feel him break free from me as Feste did, but monstrously, on a dark descent rather than airy buoyancy.

And I remember a diner on a seedy stretch of road where I could get two eggs with grits and fried potatoes and a cup of coffee for 29 cents; and the tropical rainstorms that arrived at noon and caused lawns and parking lots to steam like saunas till evening, and a Junoesque actress with orange-red hair (preserving herself for a husband) who pulled me into a bathroom in her house in Miami Beach to show me its gold fixtures, and drew me from the descent that began when I broke up with the musician-actress who wrote Zorba tunes—a descent carried to its nadir in Caliban.

We break bodies with a force we call love and others break our bodies, too. Not all can sustain the breaking and survive, but an astounding number do, considering how precarious our nature is—that evasive center we're willing to give up our lives to protect. Most of us could never sustain the breaking without the words of others. They take a hand or speak a phrase to pull us from the floor of death. Few make it alone. One can wail in insular grief for only so long before the lights go out. And none could reach out to help others bear the breaking if they didn't see in notable survivors the healed seams, the fading scars, the shattered intellect whole again, and so sense how a perfect body, the original, our example, was broken for all.

. . .

I expect to be ready to write in Urbana, but the long hot summer of a performance each evening, plus rehearsals and matinees, has me paralyzed with the actor's contagion, *Isn't-it-all-an-act?* Inbred irony. Writing is, after all, an inner expansion of acting. A fiction writer enters a spectrum of characters of varying dispositions, sometimes simultaneously, rather than one. So writers can be as rigorous poseurs as actors.

Maybe that is why, before classes begin, I am walking down a campus street carrying a cane and wearing a white suit like a Southern planter when I see the woman from my couch with curly white-blond hair come toward me beside a woman friend. I bow as she draws near, surprising her enough by this for her to stop and talk, smiling, as I lean on my cane.

Her friend, tall and plump at the stomach, dark-haired, crosses her arms and taps a foot, and as they go off I hear her say, "Never trust a guy in a white suit."

She may not trust me but is inclined to talk when I start to let classes slide. Since the day I was passed over for Hamlet all I've looked forward to are tryouts for *Richard II.* Shattuck is directing the play, once again on the Globe, and I want Richard. My writing after Miami has been a wreck, and my only consolation is the woman who so far hasn't fully taken her friend's advice. Paul nicknames her "Care."

She once studied Russian and Spanish at a private school, I learn, and now is enrolled in Russian and French—a political science major who hopes to go into government service. We sometimes glance behind or to the side, as if to catch a glimpse of what's taking place, first between us, then on the campus. Parties take place nearly every night, women are so sexually frank it makes men blush, somebody keeps providing drinks for the parties, it seems, and now and then a smell like burning leaves creeps crowd-ward from a bedroom or bath.

I move with Paul and two grad students to a house the four of us rent, across from the playing fields at Men's Old Gym. All roommates are readers, three are writers: Bovim, a chunky doctoral candidate who sips at beer all day and has written a Woolfian novel wholly in monosyllables; Sandu, an East Indian engineering student, a Sikh who enlists me, on his returns from the Laundromat, to help him wind the pastel cloths of his turbans that reach the length of our long living room into manageable rolls; Paul, me.

Two months into our stay I walk into my room at night and find Bovim sitting on my bed in boxer shorts, a quart of beer in his lap, his eyes bleared, his upper body entering the taut rockings of intoxication intended to find true center. "I don't suppose you'd want to have sex with me?" he asks, answering himself with his question. In a month he's gone.

Our new roommate, Barry Kolb, is familiar to me from a science lab—with his full orangy beard and wide smile that has the even line of dentures. He is a mathematician but a philosophy student (soon to study the Talmud in Brooklyn) and as literary as any other from his reading. I introduce him to Isaac Babel and he introduces me to the novels of Beckett, *Watt, Molloy, Malone*, and the crippled meaninglessness of each character seems to me the true state of things.

Shattuck casts me as Richard, letting me know a week ahead so I don't have to endure the wait—his words. He says a grad student who presumes he has the part will need to be talked to, but maybe the fellow is getting the drift because he has been saying that the work on his dissertation might make for hardship in the role. Shattuck asks me to scout for actors to fill holes and I persuade Paul to try out; he is cast as Bishop Carlisle, Richard's supporter at court, and now Tyner and I are truly in this together.

The owl swoops from its Quonset, wings gray as the sky but outlined in orange by a sun I can't see, as Joseph and I take advantage of the recently plowed road to down another cottonwood—or I take advantage of him, because we have to toss each chunk over a fence, wade with it through a ditch where we sink past our waists, and finally load it into the pickup, parked on the road. Not one car comes by.

Not one person is on campus, Paul notes, as we head for our house from Stan's after closing. We slip over to the quad and the white-pillared portico at the side of Women's Old Gym looks luminescent in the dark. The roof is a porch with French doors to our workshop. "Books, friend?" Tyner asks, and I picture the shiny tumble on the *Accent* shelves. The doors to the building are locked. At the portico Paul starts shinnying up a post, wiry, a former member of a tumbling team, able to do a back flip from a standing start and land on his feet, and I shinny up after, grab hold of the white railing above, as he has, and we're on the porch.

The French doors are unlocked, as we suspect, because we've stepped out here to smoke, and we stand in a midnight setting for our workshop. "Scouffas," Paul says, and I sense him as Paul must. We enter the hall to his office, both sides of it lined with review copies from decades, find cartons in the dimness, pick from the shelves what we want, and go down the stairs in stealth, out the front door, then the leg-wearying walk under that weight to our house.

I learn that Granville-Barker did not write a preface to *Richard II* and look for similar sources from an actor's view. Only a Michael Redgrave recording. Once I start memorizing the speeches, pages of poetry, I know the role is unlike any in Shakespeare, as difficult as Hamlet, exactly because of the poetry and a self-dramatization that isn't endearing. Shattuck and I agree he's not an effeminate weakling but impulsive, a chameleon diving through the rainbow in his trajectory toward black: self-destruct.

"But, oh!" Dr. Shattuck says, "how he sings his grand descriptive metaphors en route!"

Or, "He's a good and sensitive poet who's a not-so-good king." He imprisons himself, kills himself, with his self-indulgence, as I see it—by plunging headlong into his rolls of unrestrained poetry. The stages of his pilgrimage in poetry unlock portions of my mind in a reach toward— It feels infinite. In his head I learn how poetry climbs.

From my course in the metaphysical poets I see that Richard not only follows their formulations and imagery but with a formality of his own pries open areas where feeling goes bucketing down into an abyss I can barely imagine. Yet to convey this to hearers, since the states are largely internal (until his ascension in the prison soliloquy), seems close to impossible. Which is the conundrum Shakespeare proposed in this king.

I find a book I feel is helpful and pass it on to Shattuck and receive this note:

> A very good reading of the play, marking out the progress and relationships of the *ideas* in it—turning the play, nicely, into an *essay*, or even into a *novel* of an analytical sort.
>
> But the play as play remains silent. I can't hear any voices or see any movement (of bodies or of minds) . . . The text stands still like a collection of marble statues, while Traversi views each piece from all angles and writes up his judicious report. No humor, no anguish, no rage, no love, no joy. No sensual immediacy. No piss, no vinegar.

What I think I learn from Richard is the way his mind extends into distances with metaphor—as far out as I can go and then a double dose beyond, and on and on. We have two months to rehearse, with Thanksgiving and Christmas breaks, and I don't take either. But over Christmas I contract a chest cold and fever. The production opens January 9, and on the day of our first dress rehearsal I get this:

> Larry—
>
> Now don't knock yourself out today if you don't feel up to snuff. Just fool around and experiment a bit.
>
> I was remembering last night a rehearsal before Xmas—the night you were really hot with discovery—and how in the sit-upon-the-ground and within-the-hollow-crown passage you really filled the air with excitement and transfixed us with your own vital experience. Now, since Xmas, this sort of thing hasn't hap-

pened again. The last couple nights, I'd say, you have done a faithful, studious job of *imitating* yourself—going through all the motions but not creating at all, not having *any* vital experience. The result is of course slow, mechanical, as boring to me as to you.

Incidentally, it is at least 50% better—oh maybe more than that—in all sorts of technical ways—meaning, movement, articulation, business, etc, etc—all most valuable when you are finally ready to shoot. Also it would pass with the ordinary spectators as one of the very best played Shakespeare roles we've ever had.

But neither of these things is worth a hill of peapods until both you and I are excited and delighted to full capacity with your performing. Which will happen when your health is on you and you're a ball of fire in a box of coiled springs in the midst of a thunderstorm on the crest of a volcano—and calm as an icicle too . . .

Chuck

On opening night I receive, besides notes from him, cards and telegrams, and two of these set the role and the divisions in me in perspective. First, from my former inamorata, the musician-writer who is also a member of the cast, a lady-in-waiting, now truly waiting when not pursuing me in the wings, this: FOR LARRY, NOTHING MORE NEED BE SAID MY LOVE. And there it is again, that word, *love*, presumed on me in a way I mistrust, causing me to cringe. The other: FROM OUT THE SOUL IT WELLS AND THE HEARTS OF ALL WHO HEAR COMPELS. I SHALL LISTEN.

It is for her I'll do the role, I decide. I sing and cry the poetry for her approbation, my Care ("My care is loss of care, by old care done; Your care is gain of care, by new care won"), and the speeches pour with snappy springs of fiery thunder from the crest of that volcano. Halfway through, as the house clears for intermission, Shattuck appears in the wings, pale and constrained, and puts a hand on my shoulder.

"My dear boy, I don't know how you do it, but you do—more than I could've imagined. This is Richard indeed."

At a matinee, in the scene at the lists when Richard reads his proclamation, I unroll the royal scroll and below its fake calligraphy see, in Paul's artful printing, "Did you poot?" I bite down on an eruption of laughter, blowing snot over my beard, and every time I try to speak the laughter runs a cramming race to beat the words, but finally I get through the scene without dropping lines or giving up, and in the wings Shattuck is there, shaken, hands shaking. "Larry, Larry, are you all right?" he asks. "I was—" I unroll the scroll and hold it to him. "Ach!" he says. "Good God, I thought you were upchucking!"

We hear the pun the second it's out and grab each other and shake with the laughter that overtook us after Feste's final curtain call.

I wipe away the face—eyeliner, eye shadow, rouge—I'll never see again and walk out of the dressing room. Care is outside, a square slim package over her breasts. We have been listening to operas and to Khachaturian and Borodin, his *Polovetsian Dances,* as we lie with our heads touching on the living room floor. I open the note on the package, her calling card, with writing on its back—"Vive mon dieu, mon roi"—and in the package find the records of my favorite opera, *Otello*, such a lavish gift for a student I have to cough to cover my feeling.

. . .

Here, a strip of orange under a boil of dark-blue clouds so bulbous their upper reaches bump at heaven. Sunset.

2

Richard II is part of a University Arts Festival and in the spring, as concluding act, William Maxwell arrives and speaks on "The Autobiographical Novelist." At a podium I see a slight man with thinning hair in a suit and white shirt, his tie slightly askew. His thin noble face has ruler-straight eyebrows and a Scots jaw. But what strikes me is his voice, shy and tentative, with a hesitancy that seems at times the start of a stutter, and with a breathy whispery quality, as if an example of tactful gentleness resides so deep in him he will not overstep it, not with his voice.

But he is in command and when he looks over his half glasses, one of the first sets I've seen, his eyes of melting brown have the alertness of a baseball slugger's. He is a ways into his talk when he whispers with a forceful clarity that has people at attention in their seats, *The autobiographical novelist is the kind of person who can stand at his mother's coffin and notice the shape his shoes are in.* He looks up and hits me with his eyes and the statement enters me like a shock.

An uneasy amusement agitates the audience, giving me a perspective to appreciate the humor in this, but I'm not sure others see it as I do. It suggests a person who, no matter how observant, is cold-blooded, achieving detachment in the worst situation, as I do, I'm afraid, when I think of my mother's death or perform a role, watching myself watch the character being observed. Maxwell's unsparing description makes me feel less heartless, a member of anyway a portion of humanity—a trait the person at my side who is my main reason to breathe has been spared.

After his talk Maxwell stands outside the door of the room and shakes hands like a preacher at the back of a church. He looks relieved, his smile hearty. An older man ahead of me takes Maxwell's hand in both of his and keeps pumping it as he pours out praise for his books, going into detail as the priest did with Mann, and Maxwell stares straight into him, his face immobile, registering little response, turning grave, and it isn't until years down the line, when I encounter fulsome flatterers, that the look of Maxwell reappears, his somber expression as he receives this passively, aware that the man is talking through his hat. As he would put it.

I step up and before I have a word out he says, "I'm looking forward to having coffee tomorrow with you and Chuck."

We meet in the K-room in the early afternoon, its busiest hour, all booths taken, so we sit at a square table big enough only for our mugs—arms and legs angled spaghetti. "This is my Richard," Shattuck says, "the one I wrote you about. Also the Trelease winner. He writes like a sensitive young woman, if you don't mind my saying it."

"No. It means I can act." I'm uneasy myself, saying this, although they both look amused.

"Is that what you want to do?" Maxwell asks. "Write?"

"Yes."

"Then you will."

I wish my fiancée, as I've started to think of her, were present to hear this, and then I see that the black trench coat I've worn on campus from the time I enrolled has fraying cuffs, loose threads dangling.

"And can he act!" Shattuck exclaims.

I look at my hands in my lap; I've been in two roles in the workshop since Richard and now a main-stage production, but I've learned the more I enter a character, the more difficult it is to locate myself afterward, and feel so displaced by the give-and-take I sometimes wonder whose mind I'm in. Acting is easier than writing but the aftermath I do not relish. But if acting were all I did, and I were paid for it, couldn't I endure it?

"What are you working on now?" Maxwell asks.

"Oh, mostly essays. They're required—for classes, I mean—but I'm taking an independent study with George Scouffas. I plan to write a novel."

"Good."

I don't tell them and confess here for the first time: I was writing stories for others. They were friends and I wasn't paid (though I think I expected a seduction from one) except for the immediate pleasure, and then seeing the grades. The requests are desperate and I enjoy the puzzle of fitting prose to another sensibility. I seem to learn more about myself from this than from the stories I sign as mine. Saul Bellow says that when I think I'm not being autobiographical, I'm being most that way, no? Once I let fence lines sag as I do not when toeing the line of what seems actual, I enter clover I imagine is entirely other than the clover my nose is generally in, and go whole hog. So in that ecstasy the other is me.

One of the stories, for a sorority sister of my former inamorata, is about a young woman on a beach off the Chicago Loop—I'd never been to one but always intended to—and by its end I feel I know more about why the inamorata left (and then wanted to return) than I learned from her or could discern on my own. For a friend I tell the story of an unstable loner who keeps up a pretense about his need for alcohol (a breezy test of a Faulknerian voice) and at the end loses his hat. Not his head or his gloves but his hat. By the time I'm done I know this is the undercurrent threatening me and if I don't get a hold on it I could end like the protagonist, bucking over a bucket with saliva strings like his last lines to life.

Maxwell and Shattuck take their hats from a chair and stand, arranging their topcoats, and Maxwell says, "When you get to New York I want you to come and see me."

That settles it as surely as if the ticket is issued: I'm headed for New York.

Tyner has a new friend, Lo, a painter he has named after Nabokov's heroine. She spends nights at our house and one morning in the aftermath of *Richard II* asks if I will sit for her and spends hours on a pencil sketch of me, then of Care, each on separate drawing-pad sheets—grainy blowups.

"Jeez! Aren't you done yet?" Tyner asks. He can catch a person in quick strokes on a napkin, in caricatures bizarre or grotesque, lines askew.

Lo's portraits have a dimensional touch in the shading and highlights, as if a soul can be seen emerging through the features of each, and the person each was on that day is apparent, a quality both portraits retain—though they barely survived.

Lo matted them and gave them to us as early gifts. Care and I hoped to be married that summer.

We had Scouffas over for dinner and he asked to see where we worked, and I led him to my desk, which I admired, because of its neatness—a mahogany door sawed in half (I don't know where the other half was, maybe with my brother), its edges painted violet-blue.

"Oh, no," he said, and placed his fingers on the red-and-black thesaurus lying at a casual but artful angle to one side of the desktop. "You don't use a thesaurus."

"Well, sometimes."

"You shouldn't. If a word doesn't come to you naturally, it falls outside your voice."

I hide the thesaurus.

One day back from class, in the living room where both portraits are hanging, I find mine is turned upside down. Maybe a fan at a party, I think, and right it. The next day it's upside down again. I right it and next *hers* is upside down, mine missing. I find it shoved behind a bookshelf, turned toward the wall. I'm aware painters and artists get tired of their work or find it tiresome to see every day, and that's the assumption I prefer. I straighten both and go to bed and I'm half asleep when I hear footsteps go past my door. I open it and Tyner comes tiptoeing from the living room in undershorts.

"What's up?" I ask.

"You?" he says, and goes to his bedroom.

In the living room the pictures are reversed. Tyner receives a grad assistant's salary and I owe him a hundred dollars for rent, which he offered to lend me. I work different jobs at different times but I'm able only to keep up, not pay him back. I don't believe it can be that.

His brother flies out from Virginia for Christmas and in the spring a friend, a reporter for a Virginia newspaper, spends a week in the house. I walk in and find the two at the table leaning toward one another as they talk, Lo a ways back, arms crossed, an index finger to her cheek, and at the sight of me the talk stops. One day I say that a song they play, Ramblin' Jack Elliott's "San Francisco Bay" is getting on my nerves, and they play it all night. From my bedroom, as somebody lifts the needle to reset it, I hear the friend say, "Why don't you kick the worthless shit out?"

When I enrolled in an independent study with Scouffas to write a novel, I asked if he wanted to see it all. "Don't overreach," he said. "A few early chapters will satisfy me." I stay in the house over Easter vacation, alone, to draft it, and the work does not go well—a grotesque tale similar to *Watt*, with a forty-year-old man living at home, his father in a wheelchair after deciding he's tired of walking, a harridan mother with "fingers like pawls," etc.

Paul gets back and says, "How'd the novel go?"

"Only seventy pages."

"*Only seventy*? What the hell do you want!"

It comes out to sixty-some typed, a few early chapters, and I hand them in at the end of the semester. With this project the conference is all, and when I arrive Scouffas looks unwell, greenish—he's had a colon operation earlier that year, I know. He turns to his desk, where my manuscript is resting in a thin tan box that originally held the onion skin paper I type on; he turns to me, turns back and sighs, then looks over his shoulder and says, "Larry, either my judgment is failing me or this is the worst thing you've ever written."

It takes third place, a week later, in the Trelease Award, a help. I give it to Dr. Shattuck for his opinion and don't hear back. I tell all this to my fiancée, of course, when I go to meet her family. At the end of my visit she comes to meet mine, and as we ride in the car with my father we pass love notes back and forth. The meeting goes well, we believe, but when her mother arrives to take her home, my father says, "I think Larry needs to settle down before he gets married. How is he going to support a wife?"

I feel I can never forgive him for the statement. It adds to the pressures my fiancée is already feeling and for the time being she calls our wedding off.

In late summer I drive to the campus to enroll in a correspondence course in Greek tragedy. This is an attempt to lighten my load for the next year, my fifth. I've switched curriculum so often, from theater to journalism back to theater to rhetoric, where I've ended up, I have twelve hours of required courses left.

I walk to Dr. Shattuck's house, feeling iron talons in my heart, and find him in the lawn with his wife, having tea beside a flower bed in bloom. He is on his side on the grass, head propped in a hand, sleeves rolled up and collar open, a cup of tea on a saucer in the grass close; she, with her gray-white hair in a ponytail, on her stomach reading *Golden Apples*, her feet up and twiddling.

Shattuck knows how things have fallen out with my fiancée and asks about her. I feel so worn from all this I can hardly talk and pull up grass, hoping he'll mention on his own the chapters of the novel, and I sit so long in silence his wife leaves for the house as if she knows I need to talk. When he doesn't say anything, I finally ask if he's had a chance to read what I gave him. He is silent, then shows his gapped garland, as he did onstage before he took an arm to lead me though blocking he wanted, and says, "It seemed to me to suggest an awful lot of cocksucking."

His wife walks out carrying the thin tan box.

I can hardly take it, the flowers snaking up into my vision as I make it from their yard out the gate, shaken, getting out a goodbye, with not the faintest idea of how he derived what he did. All the people in the book are old and all they do is get on one anothers' nerves. Not even an intimation of, well, fellatio. Then I remember a moment when, after walking to a roadside mailbox, the protagonist sticks his hand inside, into its heat, and remembers a dream, a buffalo, nickels, an Indian with a spear coming out of his mouth. But that was classical symbolism, *symbolic*. How could he think otherwise?

Then there she is, on a corner outside the Union, with a suitcase in each hand, the one I want to marry.

I run up to her but, no, this is not for me or my sake, she is hurt in her own way, and finally says, "I had to leave home. I can't live there. I can't go to school here."

She's on her way to the city where she attended private school, with change from her piggy bank for bus fare. We spend hours together, into the dark of night, her in tears as I affirm my dedication to her, even if we can't marry now.

But the iron talons remain and the next morning I see her off on a bus for a city to the south.

When I had my first disagreement (one of the rare ones) with an editor who threatened legal action about the "next work" written into every contract, I said, "All I have to do is send a manuscript you can't possibly accept."

I would have sent the one in the thin tan box. It wasn't necessary, once we talked things through in a civilized way. You don't want to antagonize a person with the power to publish. Then again, you can't let a publisher push you around, as if you're a business dunce, or it can turn into a habit visited on every writer down the line. So writers employ agents.

All business is built on compromise but writing, which allows for a lot of editorial give-and-take to move a work nearer its original intention, is not.

3

Ruth and Laurel string lights on a jade plant in the living room and we sit and stare at one anothers' stunned faces. Only a few gifts the girls have made, and that's Christmas. I get out the electric lantern and head for the mailbox, in case somebody made it through unbeknownst to us, and the lantern's back-glow sets rings like eyeglass lenses around my eyes. These wobble so much I'm not sure I'm off balance or shaking the light. Underfoot, ice and snow at several levels, all white, buckles in upheaval, a sea.

My last semester I will not act, I tell myself, and may not write, but will be a responsible wage earner and win her that way. I write advertising copy for a Champaign television station, and then the editor of *The Daily Illini* asks me to serve as theater critic. His name is Roger Ebert, his puffy face a freshman's, though he's my age and size, in the hundred-and-thirty range, with the dolorousness of those who have to cope with too much too young. He wears a suit jacket like a badge over his youth, and perhaps I inspire the dolor, since he has to work at odd jobs in the office, exercising patience, while I type up my reviews.

Sometimes he steps into a side room, perhaps to see a film on the late show. *Thumbs up!* The *Illini* is a true daily and must be put to bed by 2:00 A.M. Often I'm not done by then and he carries what I have to the typesetter and then, one by one, carries off the pages as I finish them, and I have to say "Just a second!" and fix a sentence in pencil before he leaves.

The television station, WCIA (of all acronyms), employs three writers, all young, and an amiable competition builds among us: who can invent the zaniest copy? The in-house production for sixty-second spots starts to run to two hours and the camera crew is livid—these are for used-car dealers and insurance agents. Clients complain that the ads are too coy or "far out" but their sales climb.

The pace gets so feverish as advertisers are added that living alone, as I do now, is a pleasure. One night I sit on the couch in the apartment I rent, looking through a file folder of old poems, feeling nauseous about the next day's work. A sensation parts the back of my skull and I find I'm standing on a pitch-black plain where in a panorama I see my parents and grandparents act in silence familiar and distant scenes that have the dim strangeness of oracular dreams, and I understand that the first eight years of my life were separated from the others by the move to Illinois and lopped off for this, the bolt of cloth clopped by the arm of a cutter fluttering in its backward fall into that trunk (its secret, I see) until I break the locks and wrap myself in the bolt that keeps unfolding into broader quadrants of an amphitheater like the interior of God, where the thought is given to me, as everything drops into darkness: *This is my novel.*

I was sleeping on the subway the night the shuttle caught on fire. 42nd Street sank two feet and had to be cordoned off. My sleep was from too much beer and I woke at the end of a line, in the Far Rockaways, and found that my guitar, lying on the seat beside me, was stolen. I was so alarmed I decided to set aside my animus against older gentlemen and see William Maxwell.

Take the elevator up to the 20th floor, he tells me on the phone, but when the doors part I'm sure I've gone wrong. I'm not at the elegant *New Yorker* but in a bare corridor on an abandoned floor, so it looks—institutional green with dark scuffed linoleum. Then Maxwell comes through a door to a hall, in a suit and tie, and leads me down a dim passage to his office: a table inside the door, a schoolteacher's desk ahead, and beside it, against the wall to the right, a cream-colored couch whose fabric looks like woven silk.

Past the table to the left are windows with low sills. I go to one, partly blinded by the light at this level, and find I'm facing north: Central Park in a green swoon.

He steps up close, also looking out.

"What do you think of the view?" he asks, and his words are fraught with such meaning I can't speak.

I've sent poems from Urbana, and he passed them on to the poetry editor, Howard Moss, he says, but Moss is ill, and a letter Maxwell sent to Urbana about these complications must have missed me. I don't tell him how long I've been in the city—nearly two weeks.

He pulls a wooden chair from the table and revolves it toward his desk and by the time I sit he's in his swivel chair, hands clasped behind his head, elbows out.

"Will you be able to manage alone in the city?"

"Yes." Though the reason I've come to see him is to ask for a job at the magazine.

"In my letter I gave what I called Ciceronian advice. That awful Walt-Disneyish World's Fair may result in a kind of mass irritability, I said, which would not be helpful if you're looking for a job."

I flinch at his reading of my mind.

"So from that point of view I thought it might be worth it to think of not coming till fall. But I don't really know there will be an irritability or that it will affect things for you one way or another. Maybe we had a glimpse of that last night. Did you know the shuttle from Times Square to Grand Central went up in flames?"

"Yes."

"Anyway, you're at the stage of your life where every action, looked back on later, proves to have been destined, so whatever you do don't listen to advice from middle-aged people like me. Here, your story will walk up to you."

Outside, in Times Square, it happens—an actor from our troupe in Miami, Gil. He is on his way to visit a friend in the Village and asks me along. The friend has a bit part in a pilot that might be a help with connections. But we learn from this actor, whose name is Tony, that the pilot isn't going to make it as a regular show.

We pay too much to get into The Bitter End and sit through a comedy act, "Jim, Jake and Joan"—the woman is Joan Rivers—so we can hear a writer and folk singer from Miami perform his latest, "Tear Down the Walls!" in a break between the comedy act. Jay Robinson is in town, as I notice from posters, performing after his Miami purge on Off-Broadway, at a theater in the Village.

A few days later, again near Times Square, I run into Jude, from the theater department at Illinois, a young woman with the dazed and rattled look of a character actress. She says a friend of a friend needs somebody for a play at Hunter College, and as we talk, she says "Come along" and leads me to the Overseas Press Club, across from Bryant Park.

We ride an elevator up and walk to a door with "Attention, Inc." taped across it—a PR mailing firm with rooms so small an offset press sits over the tub in the bathroom and discarded printed pages rise to the faucets like novels gone awash. The help is mostly unemployed actors—Jude works here; we met on her lunch break—and one of the owners, Van Varner, says, "Come and go as you please, keep your own time," while typewriters clack and the press above the bathtub echoes ka-*thock*, ka-*thock!*

The job is enough for food and a room on St. Marks—a Ukrainian and Puerto Rican neighborhood, with older Italian families moving out, not the fashionable place it will be. I promised to give myself a year to write but give in to the woman at Hunter. She needs one more to make up a threesome for a recent translation of a German expressionist drama, *Three Blind Men*.

The three of us rehearse in a bare front room, eyes shut from day one, groping our way through blocking. I don't care for the play and regret giving up my plans—I vowed not to step on a stage till the year of writing was over—and on weekends I tend to drink too much, back from rehearsals. I often meet Tony in Googie's or ring his buzzer if he isn't in the bar.

Yes, Googie's. It existed then, too, but the only similarity between the original and the present place is the dark-painted toilet with a cement ledge at chest level. Tony's apartment is smack above the bar, so the nighttime music and noise drives him out anyway, and it seems we're becoming friends.

April, 1964: Time moves only as counterpoint to the isolated mind. And her inflections in time, her grace notes ad infinitum, were no longer great enough to touch him, I wrote in my notebook.

Which was my way of saying it was not *her* I missed.

On one of my first visits to Googie's I see an alum of Illinois, a person at the fringe of the people I knew, a dapper and gaunt-faced Italian, Benny, at the same time he sees me. We shake hands and say hello and it turns out he owns the bar. He only recently opened it, he says.

"Do you think he's with the Mafia?" Tony whispers when I tell him this. He is sure the actors for the pilot were hired through the Mafia, and then most of it was filmed in a restaurant owned by an elderly Italian who, Tony is convinced, had Mafia connections.

"Then how were you cast?" I ask, because his parents are Jewish; they own a jewelry store in Miami Beach, as he has told me. "Because my dad knows the guy?" He means the owner of the restaurant. *"I don't know!"* Tony puts his hands to his hair and checks his expression in the back-bar mirror. He is dramatically and comically emphatic. He can put a spin of emphasis on a question, especially after he sips through a few J&B's on the rocks; then cappuccino and bagels at the Figaro at 3:00 A.M.

Peter, Paul & Mary are on the ascent and the bartender at Googie's is Mary's husband, an attraction to women. I sometimes wander over to the Kettle of Fish or Cafe Wha? or Gerde's Folk City, hoping to bump into Bob Dylan. I meet Tom Paxton and Dave Van Ronk, who even then was singing, "Co-*caine!* runnin roun mah brain!"

One night on my way home a young man runs up, his face battered and bloody, screaming something I don't understand. Then I see he has a knife and at that moment a man with an umbrella comes by and pokes him with it. The bloodied young man swings to him but the older man holds him at bay with his umbrella tip and then a car comes by and somebody shouts from inside, "Leave him alone! Can't you see he's been hurt!"—all so bizarre I tell myself, *Never be on the streets at that hour in the morning.*

The week after I start at Attention they move to Lex and 34th, a second-floor-through in a squat building. Jude gives me the number of a grad student from Illinois, in the city to finish his doctorate, and he says with a sigh, in an accent acquired over a year in England, "Come over and I'll help you put together a résumé. That way you can, well, get a *real* job."

He lives on Pineapple Street in Brooklyn Heights, down from the St. George, an upscale neighborhood, I see, as I walk around, down a block on Willow.

We put together a résumé he feels will work at literary agencies, where I might fit in as a reader, and he suggests I first try Sterling Lord. I carry my résumé up and "have a seat," as I'm told, and page through a file folder of poems and writing samples, in case they ask, and after an hour I'm ushered into the office of a kindly-looking, older and urbane gentleman. "I had to look at the guy who put this together," he says, and shakes the résumé at me. "Where did you get it?"

"I typed it up at a friend's and had another friend run it off at the place where I work."

"Inventive, huh?"

"Inventive?"

"Nobody could do this much by twenty-two! Don't try to give me that! Go on. Get out of here."

If I were Paul Tyner I would have accomplished not only my list but then some, and could have handed the fellow a caricature I did while he talked, then started whistling Shostakovich as I did a back flip on my way out.

It was the only agency that even gave me an interview.

July, 1964: It's difficult to change the character under whose identity you first meet someone. If you were grave in their presence the first meeting, you will continue to be grave. If you were clownish, you know they expect that from you, and will be a clown. Of course you'll feel other emotions but they'll all be under the auspices of the dominant one. . .

Only a week into rehearsal, one of the three blind men says he can't sacrifice his job for this role, as if he could sacrifice it for another, and I can sympathize with him. He's a taxi driver. A friend of his knows somebody, he says, who would fit the part—in this poor play that seems to be cast by friends of friends. The substitute arrives the night after the taxi driver drops out, and I learn he's only nineteen. His name is Robert De Niro.

His father's name is Robert, he says, and asks us to call him Bobby or Bob. The other of our blind trio, Syl, a morose Italian, says Bobby is too rough. As a diamond, yes, maybe, but also just rough. He seems oblivious to an unwritten rule of the stage, which is never, no, never manhandle a fellow actor, especially his or her body, but do only what you must for things to look good out front, so you don't crimp an actor's style or ability to act. But this is pretty much a Marquis of Queensbury rule, ignored when people get heated up. Bob grabs hold so hard in our blind grapples he is, according to Syl, out of control, wholly internal, with no sense of ensemble work.

I'm amused by the cockfight between the two. Syl wears worn clothes and a black leather jacket and Bob usually arrives in a dress shirt and suit jacket. He's out to please, with an easy, unselfconscious smile, which can shift into one of such abandon it draws his hairline back—the grin of a young man secure in his new maturity. He *seems* eager to please but is a lightning rod attracting the emotions of any observers in the room on a given evening, pulling in all that and everything Syl and I give him and returning it at a higher level, volatile, a presence to work with or to back from, not a performer of mere skills.

Once he is done for the night and out on the street, caught off guard, he looks entirely like himself yet different each time—one evening a bashful altar boy, his collar humped up from his jacket so engagingly I want to reach out and straighten, and do. Then he's a sallow sensualist, Gallic, or as mussed and plug-ugly as the pug in *Raging Bull*. You see this chameleon makeup in *Ronin*, one of his more subtle roles, where for the first scene he looks like an eponymous Parisian from any back rue, and then as he asserts his questioning nature, his features and expression alter to American, the person playing the earlier role, until he appears to be Joe down the block.

For his role at Hunter, he watches blind people navigate streets, how they walk, and goes to a home and talks to some, the kind of research I like. I take him aside and ask where he's acted. "I did mostly scenes." Bang. His unrestrained smile, with no hint of embarrassment, sets the lines of an oldster into his cheeks.

"Scenes? Where?"

"I'm studying at Stella Adler's." He gives me the swift satisfied grin, both eyes turning blank, that reviewers later call "goofy." He can also focus one eye directly on you with the same grin while the other goes blank. "You know, where Brando went. *Mahlon*." Trying a new voice, his lips drawing toward his rising hairline.

It's difficult talking to him, because any question moves him to another quadrant of character, if not a new character altogether—that chameleon nature. His skin has the translucence of youth and he can turn his face so that one of the two moles on it—the larger high on his right cheek, the smaller just back of his left eye—sets the focus of his features in a different cast. Again another person.

But a chameleon that is trustworthy; his face glows with the inner light and loyalty seen only in saints—the stunned ecstasy of the conquistador in *Mission*.

. . .

Adler began her lectures, as a recent transcription has it, by saying, "This is not a course in 'drama.' It is a course in opening up the vastness in you as a human being, in all your aspects, to understand your place more than you do—not to be led by the Bible or anything else but the truth of modern life as given to you by certain genius-authors in the theater who can make you into something tremendous. That is why you are here."

His suit jacket is usually dark; one tweedy sport jacket is fraying at its elbows and edges, his dress shirt is always open at the collar; slacks, street shoes. Acting for him, more than anybody I've worked with, is a religion. Grail. "This other stuff is technicalities," he says. But it's a religion with roots in an area deeper than Adler's nostrums: *family artistry.*

Nineteen is so distant from twenty-one, to a person past the dividing line at twenty-two, as I am, that it seems the arena of high school. I try not to be snobbish about that and appreciate his street smarts and sense of the city. But I feel his voice is too frail to bear up under an extended engagement onstage, with its high thin edge that means no diaphragm support, besides its tendency to climb higher in the pattern of many actors when excited, a pleading plaint. And its thin edge is blurred by a smiling fuzziness. Any outstanding talent isn't apparent in him, not to me, in my perhaps priggish sense of classical training and its techniques. But something better is apparent.

It would settle here; acting to him wasn't quite grail. He considered setting it aside for a while, perhaps for good, if he had to, before he got the part of the blind man, and he was just back in the city after months away, in Europe. Syl subsided and did, like the duo of the trio, the best he could. The play got up off its feet and onstage and we did our blindness in the black night of the blind and received a few reasonable reviews. But the play itself was not the kind (a clunky precursor to Beckett) that anybody was going to write home about.

Perhaps Bob did write to his father in France about it.

What Do You Tell the Folks Back Home?

I am out of money, except for the dozen hours a week I work at Attention, worse off than a student. I send a note asking to see Maxwell and hear right back, "I'm running off schedule this week, and you might not find me in. If your job permits it, would you like to have a quick lunch with me this coming Thursday? If you could be at the office at 12:30, that would be best for me."

"Going to see Maxwell again?" Van Varner asks. His favorite writers are Hortense Calisher and Muriel Spark, so he says, but he knows Maxwell's novels and stories, and has met him. "I live down the block from Howard Moss," he said once.

"Edward Albee is my neighbor." He puts on a prim face of fake concern. "I'm a kind of writer, too, you know!"

He is so energetic a jumpiness often overtakes him. "How now!" he'll cry, and spin to the postage meter, rapidly feeding a stack of letters through, baring his teeth in a fixed grin, jerking glances at the unemployed actors sitting around in their depression. His red-sandy hair, like his exuberance going up in flame, is erratically mussed, his tie looks stuffed or crooked in his loosely bulging button-down collar or drags over one side of his suit from his hurry—now tapping first one foot then the other as he works the postage meter, his pouter-pigeon chest pulling at the buttons of his shirt.

"Yes, I write!" he adds generally, to any who might be listening. "I do, I do! For *Guideposts*! The truthful little mag of everyday hope!"

Attention handles *Guideposts'* publicity mailing and is in charge of their exhibit at the World's Fair, and Van is indeed on the staff, a contributing editor. The Fair, far out in Queens beside the gaudy stadium of the comic team meant to compensate for the loss of the Dodgers, is not so disruptive after all, as I find one weekend when I take the subway out to deliver brochures—tied by twine, with a wire handle in a cardboard roller twisted into the twine at top. I find myself greeted by Norman Vincent Peale, a cherubic version of the photo of Oswald Spengler on *The Decline of the West*, one of *her* favorite books, often in her arms in Urbana, along with Theodor Reik's *Of Love and Lust*.

The daughter of Leonard LeSourd, editor of *Guideposts*, the second husband of Catherine Marshall (of *A Man Called Peter*), works at Attention that summer, now and then helping at the *Guideposts* counter in the Billy Graham pavilion, a dark beauty with a settled regard on a realm greater than herself, able to look straight at you but not with the look that invites you in, and can swing her body toward you, with plenty to swing, yet not offer it. I'm able to talk to her only in the joking banter of Van, aware she's unapproachable by such as my pig claws, or so I see and recognize now, God forgive, nor did I ever press an approach, God forbid.

Maxwell and I take an elevator down from his office and he hails a cab and says, "Fifty-ninth at this corner of the Park." He is carrying a brown paper bag and we sit on a bench up a slope from the lagoon and eat pieces of chicken that come wrapped in wax paper, as if prepared at home. I gulp it, so desperate to tell him I need a job I finally can't talk. He mentions writers he thinks I should read, and I may miss one, crammed so full of the question I must ask. Soon we're in a cab and back to his office and he's staring at me, hands clasped behind his neck. I tell him I'm trying to write a story, but—

"Good. Whatever you want with your whole heart to do, you will. Chuck says you're a writer, and I'm sure you're the kind only you can be. It's written all over your face. It's the half-formed and half-hearted desires and ambitions that go unrealized. But you know what you want and so that's how it will turn out. No force in the world, no person living or dead, can alter that.

"If I seem to be meddling too much, you have to say so. You have to make allowance for my concern for people—that they are not cold, that they are not hungry, that they have a roof over their head, that they don't have to bear alone a terrible weight on their heart, and on and on. I can meddle in that way and you'll have to tell me when to stop because I do not in any way want to come between you and your will to be a writer. Do you have money enough to live on?"

"Yes." It shames me to think I was planning to ask for a job when by his word he has made me a writer.

JULY. I believe I have found an important part of my subject matter in the use of actors & theater. My concern for the real over the artificial, the symbolic over the formless, the eternal over man, has, with this subject matter, a tremendous opportunity to speak without seeming contrived or didactic. Besides, I am *theater.*

It's good to give yourself advice when young though tough to take. One morning I'm walking the streets again, on a roundabout route toward my place, when a tall person goes past the lit windows of a bakery, in the opposite direction, with a cut across a cheek that looks fresh, a woman, I realize now, breasts, and turn. My first impression was of a man in jeans and a checked shirt, bronze-red hair clipped close in clinging curls. Something is wrong. I catch up with her and try to talk but her eyes are glazed beyond drugs, focused on an internal terror.

I get her to an overnight diner and buy her coffee but she lets it sit as she stares away with her terrible look. A young man moves over from another table and is so pushy in his attempts to chum up I feel he's trying to make a third, in the way that the bad in the city goes downhill to worse. "Leave her alone," I say. "Can't you see she's been hurt?"

That seems to reach her. She lets me ride with her on the subway and walk at her side to her place in the seventies, just off Needle Park, as it's called.

A presence of evil bulges through her room. A green light bulb is fastened upright to the only table, for heating smack. Her friend was a user, she says, speaking at last. "That's him." A framed picture stands beside the green bulb—a head-and-shoulders shot of a light-skinned black in Army uniform. "The Army fucked him," she says. "Then they made him leave before his hitch was up and screwed him worse." He wanted to be a writer, she says. She did, too, as her story emerges. A trio did and all three ended in a relationship to one another and to books, the person in the picture, Gene, on an unproductive tangent, driving a delivery van for Bookazine.

She met him in high school, where she was the editor of the school's literary magazine and made it, she says, only with women. She picked out the beauty of the school and won her as her lover. Then Gene stole the beauty from her. So she stole Gene from the beauty. She lights up. She smokes cherry blend tobacco in a curved-stem pipe and calls herself Dylano. She has Thomas's liverish lips, perfect for a pipe, his fleshy nose and sloped-back chin. Her favorite writer, she says, is Janet Frame. She works for the publisher George Braziller.

Gene was on smack when he went into the Army, she says, and at Fort Dix he got caught and was kicked out. He tried everything to break out of it, and a priest or pastor even found him a room upstairs in a cathedral. He swung from the beauty to Dylano and back. "But that femme never loved him," she says. "I did." He got worse, she says, so bad he couldn't live with himself and the drugs. A month ago he bought a pistol and shot himself in the cathedral.

"And, oh! he's been trying to come through the ceiling! He's trying to bring people with him. He was tonight! I fell down the stairs trying to get out! I cut myself then!"

The bulging presences shift and stir and in the midst of it I think, *This is my first novel*. But I've heard only the first installment; a while back a young woman was murdered in Central Park. "Have you heard of Philip Wylie?" Yes, I say, with a grip of fear. He is the author of a popular book, *Generation of Vipers*, a title taken from Jesus' excoriation of the Pharisees, and it was his daughter who was killed, in a brutal murder, I know, because its details and his outrage made the front pages.

An upstairs neighbor of Dylano's with a German boxer, a weirdo, she says, came running up the steps the day Wylie's daughter was killed, just as Dylano stepped out her door. There he stood, his T-shirt and pants spattered with blood. He grinned and said, "My dog is having her period."

"How much blood does a dog have in her period?" she cries. "Not that much! I know, I know! I haven't seen the bastard since, but I'm afraid I will! I'm afraid that'll be it! I'm afraid I'll open the door and he'll be standing there with his shit-eating grin! I'm afraid he'll be coming through the ceiling!"

I glance toward the door, planning my escape, when she tells me she is trying to move out. She has rented a house in Brighton, where she grew up, not far from her parents' place. She wants to move this week, at least partly, but doesn't know how to, can't; the house is bare, not a bed to sleep on, all her stuff here.

"Get your sheets and the rest," I say. "I'll get the mattress."

It's in a small side room, a cubicle above the entry, packed with greater malevolence. She doesn't believe me but I grab the double mattress, balance it on my head, and ask her to open the door. "Get blankets," I say, and we're out on the street, heading for the subway to Brighton Beach.

At one stop we have to wait for the express an hour and a transit-authority cop (so she whispers) walks up to where I see his shoes and says, "Hey, you under the mattress, are you straight?" I have to slide it off to assure him I am, my neck shooting up like a goose's with the weight gone, and at the secretive amusement in her I see the Augustus John portrait of young Dylan Thomas incarnate, his scared askance look, the same ringlet curls, hers orange.

In Brighton I stare, unbelieving, at the miniature yards and gardens carefully kept, as if the weight of the mattress has transformed me, as it feels, into Gulliver. In the shine of a moon the flower beds look seismic in their similarity, about to rise and sing, yet preserved, stilled, varnished, and then I realize they are like an aerial view of fields in their precision, and that I've been in the city for months without seeing soil. I pay such attention to the yards, one after the other after the other, I have no sense of where we're going or how we got there.

Bob gave me his phone number and address after the play and expected me to call, I know, because we run into each other in the Village and I can see he's unhappy. "Come *on*, come see me," he says. "I heard about a movie they're having tryouts for. Why don't you come with?"

He gives me his phone number again and the next day, at the tryout, we walk into the bare front room of a ground-floor apartment. Two older men with beards and berets like fifties beatniks sit on a flattened bicycle carton on the floor, and one asks us to sit on another carton, across the room from them, as if it's a couch, and improvise a scene. I'm not sure he's serious. He gives an idea of what he wants and Bob starts up, struggling, but soon he's talking through his smile, and to his questions—I'm that stiff—out come one-liners, ironic, that stop things dead.

When we're leaving one of the beatniks stands and says, as if in explanation, "We're still getting set up."

In the street Bob turns on me. "Hey, what's wrong? What's the *matter*? You weren't helping in there."

"The whole setup, I don't know, it froze me up."

"Don't look to the outside. See your internal role."

"I know. I can't."

"Come on over."

He lives on Irving Place, a few doors down from Paul and Jimmy's, kitty-corner from a haunt of O. Henry's, in a renovated building—a narrow studio with shining parquet. A round oak table, rubbed to the gloss of a family heirloom, sits at its center, two wood chairs near, a cot to one side. That's it. He sets a bottle of wine on the table with two glasses and pours. *He stares at his glass on the table, and a far-reaching solitude in him takes on the shape of the room. I'm permitted to observe entire solitude. He'll achieve what he wants with this ability to give himself over entirely to himself, and that is the way to know what anybody will amount to: by the quality of unselfconscious silence they keep. The room is purged of everything but his mood, an instrument struck—the accordion gate across the window like a snag in his thought. His eyes remain on the wine, his face gold, and the table seems to revolve until the bottle of wine is back in front of me, empty.*

Some nights we sit in Jimmy's until closing. We are at the fringe, the edge, and that's our attraction—so obsessed with inner concerns we would be called, if caught in a mental ward, mad. We run on the energy of the inarticulate, inarticulate even about ourselves, I as much as he or more, because the search for words in a beginning writer is as elusive as the search for physical expression. I'm able to talk about acting easily, because I'm out of it, as I see it, and he talks about the books he goes through for a speed reading course. "The same one JFK took," he says. He wants to read scripts quick and, even better, be well read, and he always has a book or two in a battered European briefcase with two raggedy leather straps.

"*Who?*" he asks, meaning where to start.

"Gide."

He makes a statement about Dostoyevsky and I sit in silence and consider it, then I mention a scene from a Bergman movie—a kind of conversational chess. Some nights we use only facial expressions, a gesture, to indicate what we intend to say but don't, although the enactments become so intricate they're a form of conversation, until one or the other cracks up. Some nights, with enough to drink, I'm a notch near eloquent while he answers only in monosyllables; the next time we meet it's the other way around. Chess in the midst of revolving mirrors with characters added at quadrants with a turn of his jaw.

Tony and I talk about techniques or skills, or I listen as he talks about these or the details of an evening at Berghof's, where he studies, or simply the details of his day. Bob searches for a world in the round. What works for a role works, I figure, however you go about it, as long as you can *react to the person onstage*, because that opens your emotions to them. This is different each time because each person and situation is different.

The talks with Tony take up a part of most of my days, in the innocence of his need to be liked—the way he'll run his hands over his hair and emerge changed and say, "Don't I look handsome tonight?" in absolute seriousness. I understand he's too intellectual to be the aggressor an actor in the city has to be. He's simply *too nice*.

Tony and Bob do not get along.

One weekend I walk to Dylano's in the seventies, where she is staying weekdays until the end of the month, and she looks changed, her skin radiant, as if she has scrubbed herself into the shining morning face of Jacques, her inset yet protuberant eyes made up, wearing a skirt and blouse.

"I just got back from work," she says. She has let me in after checking the stairs, and now she leans against her door, shutting it with a clong, and says, "I knew it was you, after I've been so sure all week I'd never see you again."

It seems we enter a psychic state, or did the first night, and happenstance is a habit. One of the next times I stop by, unexpectedly again, on a day when she's usually at work (filling orders for Braziller, I learn), she is in. But she gives me an odd look at the door, twisting her bronze-black eyebrows, with the glaze in her eyes I saw that first night.

A narrow platform runs under the windows at the front of her room (with only the side cubicle, above the entry and mailboxes, for sleeping)—the lip of a stage that caused me to speculate in a distracted way, that first haunted night, whether a mostly spoken play could be performed there, and this afternoon a young man is stretched out on the platform, his back against one of the deep window wells.

I look at her, he looks at me, grins, and says, "Her new boyfriend?"

I'm not sure which of us he means but his question has a sear of malice. Then he lifts a saucepan to his mouth, its handle sticking past an ear, and slurps liquid from it.

He notices how I'm watching and holds it out by the handle, a host. "Kool Aid," he says. "*Cherry*. Want some?"

But he's on his feet on the floor quick as light on a panther, and as quick at the door and out. I start to ask Dylano if he is a boyfriend but before I get anything out, she puts a finger to her lips. Then she walks over and whispers in my ear. "It's the creep, the guy who said his dog had a period."

Then in an even fainter voice, "Help me."

The buzzer goes off and we jump. I see a knife on the sink counter across the room, behind a folding room divider, and calculate how long it will take to grab it. Then a knock rattles the door, though she hasn't buzzed anybody in, and settles to a gentle tap. "He knows it's us," she whispers. "What can we do?"

The voice of a woman says from outside, "Hey, I know you're there." A key like a claw clacks at the lock and a woman enters, petite and shapely and with long dark hair, and I know it's Dylano's beauty. Brett. She has Dylano's trust, she has her key, it's her first time back since Gene killed himself (as it comes out in their talk), and I can't help seeing her as accomplice to the creep who seems truly homicidal. I mean who drinks Kool Aid from a saucepan?

Dylano has arranged for Brett and a new boyfriend to rent the back rooms of the house *she* is renting in Brighton, and Brett has come here to talk about that, I learn, but she hardly speaks, mostly preens and stares, lifting strands of hair from her shoulders and pulling them behind her broad-brimmed white hat. Dylano tells her sotto voce about the visit of the creep, whom Brett has heard about, and as soon as she hears she's gone with his speed.

"Yes, that's her," Dylano says. "Ma femme."

"So why has she got a boyfriend, and why did she keep staring at me like that?"

"She's probably taking the guy, his name is Stuart, the brightest kid from our school, for a ride, you know, so she can live with an 'intellect.' Hey, she liked my pipe! She and I are on rebound, boy, and she's always interested in who's with me—that old jealousy from way back. Yeah, don't you wish she was interested! I tell you she's such a femme only a woman can get close to her. I was too rough."

Tiger Balm on her toilet top, a concoction called Tiger's Milk she drinks, and for food mostly fruit and yogurt or wonton or matzo-ball soup from a can, keeping in her bohemian isolation a kosher house. And for all her wayward looks and behavior, as seen from the outside, she is tender toward her parents, solicitous in visits and dinners every weekend, now that she's well again, an enigma. Or as the minstrel of the time, the Dylan to her Dylano, was to sing, "Brighton girls are like the moon!"

Out a window of her house I see Brett at the edge of the backyard, wearing jodhpurs in the simmering city summer heat and slashing at weeds with a stick, over and over, as if taking this out on the fellow behind the door to the rooms rented from Dylano.

"What's with her?" I ask.

She comes over, close, but keeps from touching me—her way. "Oh, she always has to wear some damn costume."

Bob's parents don't live with each other but aren't divorced, and he tries to honor and appease, even reconcile them, in the way of an only child. He is really Robert De Niro, Jr., and city friends call him Junior or Bobby. He portions his time between his parents, which is why he is only recently back from France, where he lived in the French countryside with his father—a painter with followers in the city and a gallery that handles his work.

Back here, Bob visits his mother every night, and sometimes I walk along. He's been writing his father, he says, trying to persuade him to return to the city, for the sake of his career. It's a surprise to find his mother runs a printing firm like Attention, and she and I talk shop. With her presses and photo-production ability Bob is able to update his portfolio weekly, adding photos or a slant to a résumé to match a role. We walk over from Jimmy's or Bob's apartment, sometimes late, and one night he says, "I can't keep doing this. I have to get a bigger place, a bigger apartment, closer to Mom." She hands him money, cash in bills big to me, twenties, when he tells her he's low—even for us to go back to Jimmy's, which he'll say with a smile we intend to do.

She lives in a back room of her business. Before Bob went to France, he lived there, and expected to afterward, he says, but— He compresses his lips, his eyes widen, and he hunches his shoulders, palms out, an amused beggar's parody: *he got interested in women*. When he's unhappy his parents come up, and he turns on himself as if the responsibility he takes for keeping in touch with them, in order to keep them in touch, identifies him as the reason they're apart.

For a while he hoped to follow the work of his Grandpa De Niro, a man he praises—a city policeman, retired. Bob even enrolled at the Delahanty Institute, a police academy that leads to the NYPD, but dropped out when he was cast in *Three Blind Men*. He never went back. In his apartment he has most of a police uniform and a nightstick, or "bat," as users call it, he tells me.

"Dela-*hunt*-y!" he'll grunt, to indicate what he left behind.

His mother had been a painter, too (it seems entirely in the past tense), and for a while also lived in France—the entire family did, when Bob was younger, and a talent he lists on his résumé is the ability to speak French. His mother was in the circle of friends of Peggy Guggenheim, who bought her work and hung it in her houses.

Bob's French girl, when I finally meet her at a cast party for the play, looks older than I imagined, with sharp grainy eyes that conceal something I've seen before and suspect Bob doesn't know about, though I couldn't define what that is. I can be that dense and arrogant in my take on others.

A new addition appears at Tony's, Tess, a young woman, an artist with promise, who has graduated from high school and is enrolled at Pratt Institute. Her mother, again from Miami, trusts Tony to take care of her daughter—even to indulge as he wishes, as it's understood between them, as I understand it, an arrangement I can't fathom. Tess stays with him and sleeps with him on weekends and he buys her clothes and takes her to the Village Vanguard and the Five Spot for live jazz. I find all that deedle-deeing tiring, I tell them, and so cut myself off from being invited.

Dylano likes Chaplin and I say I never understood the attraction. I mean, *I'm* a working guy, and I don't find him accurate. So she says, All right, we'll go see. So to the Little Fox where the peanuts are free and the film exorbitant, by Village standards, we go and sit though Chaplin shorts that keep her in stitches. After a while it's so much the same to me, the walk and gestures, the facial expressions, it's clear he didn't train in any formal way and is making it up on the spot, with not many more than a dozen reactions in his repertoire.

Afterward I say you can't compare Chaplin (as a director anyway, with his technical flaws) to an artist like Bergman, and she says, "Bergman, crap. His stuff is so sanctimonious it's like— What is he, a Lutheran, right, as in Martin Luther?"

We go to Tony's, with Tess present, and I introduce them all and watch Tony wince when Dylano says about yet another popular British singer, Petula Clark, of "Downtown," Tony's favorite song, "That broad's got nothin!"

After an hour of Dylano silence she crawls out an open window onto Tony's fire escape, sits cross-legged on its platform, and lights up her pipe. "Neat!" Tony whispers.

"About Gide," she says from her perch, answering an earlier statement. "Skip him for the big artillery—Genet."

Aug, 22, 1964: Let me, then, for interest and for future reference, put down my expenses . . .

70¢ for a "Hamburger Plate," heaped with fries and with a thimble of slaw and a coke.

.25—tip for the waiter

.35—a pack of cigarettes (Tareytons),

.20—ice cream cone (black raspberry),

1.00—to a beggar, the same one (and I'm not letting my need for penance sway me) I passed up several days ago. Now he is near death. He can only talk in a breathless rasp, has hardly any coordination (he sat all the time I talked to him) and is nearly blind. He is dirtier now and his sooty face is streaked with tears. His last words, "I'll say a prayer for you, I will. Excuse me, I can't talk"—holding his throat with his hand—"I'll say a prayer for you."

"Please do."

What made him so breakable, broken?

.15—*ice cream (soft ice milk)*

3.00—*two tickets to "Nickelodeon Nights," old-time movies, at the Little Fox. The Chaplin films terribly edited and after the second, repetitive, mannered and dull. Lon Chaney superb in "Phantom of the Opera." His precise stylized gesture was believable and moving and must be akin to what the Greeks did . . .*

.50—*two orange drinks at the theater (amazingly exorbitant—it came in those little square paper containers—and was paid for by the free peanuts)*

.20—*slice of pizza*

.45—*subway fares (three rides)*

Some weekends I give all my money away to street bums and have to walk to Attention, weak-legged and trembling from no food, and often on those days, near noon, the other partner at Attention, Harold, will stick his head into the back room and say, "Who wants to gets us burgers from Prexy's?"—a joint down the street. "I'm buying." Harold is the chief typist or typesetter for the press run by Don, who is from Brooklyn and has a gift of mimicry and memory that enables him to reel off dozens of Burns' poems in a Scots brogue, although he isn't one of the unemployed actors. He's better.

At the end of a penniless day I have to ask for a cash advance from Austra-at-the-front-desk, a pursekeeper who is scrupulous never to lend as much as I earned the week before. She is executive and receptionist and secretary and gatekeeper, a single and saintly woman who might have been a nun but turned herself over to Van and Harold, or rather to their business, as servant. Both are bright but don't have a business sense of organization. Van goes off on dancing creative flights and Harold, who can rattle away on a Selectric at eighty words a minute, a graduate of Pace, is a task finisher. His wife, who some days appears wheeling a baby in a stroller, with another child walking beside, gripping its rail, has a severe look but in conversation is gracious, solicitous, and I imagine it is her money that funds Attention and keeps her face in its concentrated grip. She speaks mostly to Austra and Harold.

IBM flies Tyner into New York, to interview for a job at Poughkeepsie, and he shows up at my room. "*Jeez!*" he says. "Can't you find a better place?" The same weekend three further actors from Miami appear, in touch through Tony's grapevine, and Tyner, who has left Lo in Urbana with their infant son, has hooked up with a grandson of Igor Stravinsky, and young Stravinsky has the key to a carriage house on Washington Mews.

The Miami actors and Tyner and I assemble there for a party. I can't locate Bob and Tony isn't present or leaves early, probably on Tess's account, due to the atmosphere. Paul is carrying a fifth everywhere and slugging from it, and I, too, feel looped, entering the auditory wonders of a potent weed Paul dishes out. A balcony runs around a central room, with bedrooms off it, setting the ceiling at two stories. Antiques everywhere, a grand piano to one side of the room.

Stravinsky sits at it and plays and an actress from Miami, a wise beauty I admired from a distance, appears on the balcony and starts improvising in blank verse.

I answer with my own and then, finding a character who has the knack of it better, I go up the stairs toward her, pausing to versify at levels on the way, and once I'm close to her I see, as clearly as if she's whispered it in my ear, that she admires me—a striking woman from Miami Beach so far beyond me I never considered talking to her. This seems the pattern of my life, discovering that a woman I wouldn't approach has been waiting. Her irises open, sculpted lashes trembling.

The piano stops, absolute silence, and then a torrent from Tyner in mock Shakespeare, the kind he recited during *Richard* and has written a sample of in my playbill:

Fear not, my lord. That cower that made you ping
Hath cower to peep you ping in spite of awl.
Carlisle called and recalled. P.S. Did you?

Meaning poot. Now he's mounting the steps reeling this punning gibberish off as if he has reams of it, then he takes the woman's hand, and says, "Dance, dear?"

While I think, No, don't, because I know if she walks away that's it. Which she does, looking back, eyes fixed on me, then down the stairs at his side, just as somebody finds a record or the right station, filling the room with rollicking music. The two start dancing in the center of the room, and when the music picks up Paul goes catapulting backward in a reverse flip, and lands on his feet, but harder on one side, leaning that way, hopping on the foot as if he'll fall, then is dancing again.

In a hall off the room I run into Bill Maxwell, every early novel lined at the end of a bookcase shelf. I pull out *They Came Like Swallows* and open it. At the page where I begin a stir of feeling causes an unstated part of the book to tremble under its prose, drawing the story around me, front to back, and I sob at the simplicity of it, the complexity of death, never conquered, never resolved, not in this life. Then I hear a crash so loud I'm afraid it means death, and I have a vision of Paul tumbling from the second-story balcony.

I hear him cry, "Jesus God!" and go in a grass-fogged run of slow motion to the main room. *"Jeeez-uz God!"*

He's fallen down the upper flight of stairs, flipped and hit a half-moon table, an antique, on the first landing. It's spread flat as a mat of splintered matches. His liquor bottle is smashed. "Holy balls, Stravinsky!" he cries, "you got some *glue*? Because if I can fix my bottle here, I won't have to lick this up."

Everybody deflates, the party levels off, then Paul picks up a plastic bag and gets a joint going around so fast everybody reaches *soar*. He may even say the word. I see a specter like the ghost of George Washington in a tumble from his perch atop the arch at Washington Square a block away, powdered wig and wooden teeth, come thumping dizzily down the stairs, malevolent, as if he's arrived through the ceiling, as Dylano said, and know I have to call her. I find the phone and when she comes on she's hysterical, then finally gets out, "I was *raped!*"

She and Brett were both raped, she says, by the same guy, Brett's fault, because Brett wanted to score so bad that when the black dude in Washington Square said he had a treat in his room, Brett had to try it, and they went to his awful filthy place, and what it was was a knife, and as he did one he threatened to kill her if the other bothered, taking Dylano first.

I have to get out, I say through the noise, and once out the door, in light, Stravinsky is at my side, as if he knows what's happened. First I walk fast, trying to walk off the effects of all I've had, then what I've seen, then heard, and he keeps at my side, saying he doesn't know how to handle what Tyner has done and how it means *he'll* catch hell, or worse, on and on, a comfort merely to hear his voice and to know he needs help and is talking as if I can supply it.

And just then, as we work our way around a queue crowding a sidewalk, a voice says, "And what's *this?*" A professor from the theater department at Illinois, Dr. Hewitt, stuffy compared to Shattuck, waiting in line with his wife for tickets to a matinee at the theater with its posters for Jay Robinson. I must be stained from the way he looks me up and down. Then he gives Stravinsky the once-over and says, "Boys, what do I tell the folks back home in Urbana?"

I stare at a bag of brewer's yeast and hear her sobs, my shirt front wet from her tears. She wants me to hold her and her clinging grip on me, I confess, is unsettling, because it's true, as every woman knows: there's a stigma attached to it that men can't take. They figure there must be *something* she could have done, and I fall straight into that pit.

In the back rooms, behind the door always kept shut and locked, I hear Brett's alto wails.

I finish a story, "Five Letters," about the correspondence a character reads and responds to over a night, with interspersed passages that tell another story and contradict his letters, Gidean, and take it to Maxwell. I've mentioned Gide before—my fascination with the *Journals*. He asks if I've read Colette, then brings me up to date on the difficulties-of-learning-to-play-the-piano-after-fifty, as he's been trying to do and has talked about before, then says, "I looked at Gide's *Journals* and the more I hear about his piano playing, the more I'm led to believe he could barely play a lick on it and was making half that up."

I feel the heat of this and he looks startled, his lips parted, and then we start laughing at the same moment—those gleeful seizures of his that seem will never stop.

"Have you read *Rabbit, Run?*" he asks, wiping at the corner of each eye with a little finger. I tell him I started it once.

He takes me down to a bookstore on the street and puts a copy in my hand.

The next week *My Mother's House & Sido*, pulled from his library, arrives in the mail, with a note saying this is closer to what I do. I'm so taken by Colette I'm reluctant to return the book and when I do he says, as if he's read me out again, "No, keep it. I intended it for you."

Aug 11, St Marks. Absurd as it may sound, I believe that the isolation of my protagonists in a high place (attic, hay mow, five-story room) is an expression of their desire to get nearer a mystic source: God, if you will.

One late afternoon I go up to Maxwell's office after a couple of beers and as I start to leave, he says, "I'm uneasy." He puts on his hat and scarf and coat and asks if he may walk me to the subway, then asks where I'm going. "Tony's," I say. He accompanies me on the subway and walks me there. The next time he sees me he says, about Tony, "I won't worry about you anymore. You're in good hands."

Aug 13. Long talk tonight with Bob; his troubles with the French girl. But every topic we touched upon is mentioned in this notebook, especially in the painful silences at its beginning. Which were the weeks when I was in the city alone, feeling banished from Care.

The temptation of large, rococo words. Shun everything that is not precise, honest, and unobtrusive. Allow the thought to carry the words with it; don't limit it with rhetoric or pedantry. I write all this tonight in haste, guilty that I didn't keep at my story.

In this story I include Shattuck and Scouffas, melded into a character called Serafis, and Tyner and my fiancée, and my parents, and more, pouring enough for two novels into it. I ask Van's permission to stay at Attention after hours, to use a Selectric to type it, and get so tired I lie down on a piece of cardboard in the press room, and wake to screaming. Austra has come in early and seen my legs and is sure a vagrant has broken into the offices and died.

Van asks me to stay after work and I think, *This is it.* When everybody is gone from the office he grabs some typed pages and comes in a dance to the sorting table where I sit.

"Will you help me with this?"

It's a first-person piece for *Guideposts*, about a young man who returns home and "for the first time" recognizes the worth of his parents; it isn't bad, the writing is clear, and I'm pleased he has asked me to help. "Is this about you? Are you from Kansas?"

"Oh, no! I'm a *Suthen* boy. Vah*gin*ya."

"This never happened."

"Well, *kind* of. We don't always get such wonderful stories and have to rewrite or fill in with our own. You know about editors and *quality*, no? What do you think?"

"It's not bad."

"Not a little forced at the end when I use italics?"

"I was going to say you should drop those and simplify the language right there, so it's more like the character's thoughts at that age. Then it'll be better."

"Thank you. *Thank* you!" He grasps the sides of my head in his hands and kisses my hair. "If there is ever any favor I can do, ask, even to staying in my extraordinary bachelor's pad, if you like—while I'm on vacation, of course. Usually in July, by the way."

I do not wonder if he's gay, nor is that a criterion of judgment, either way, since I worked with theater people for years when "gay" was recherché and know they're usually more circumspect in their approach (partly because of the era, perhaps) than the Don Juans and sybarites.

On a visit to deliver "Five Letters" I learn from Maxwell that he worked with Nabokov on *The Defense*—at the time appearing in episodes in the magazine—and since Nabokov is the hottest thing since fry bread, so I feel, I ask, "What did he *say?*"

"I hardly remember. It was one of those talks where with not so many words you seem to understand the other. After our work I walked him back to his hotel and he asked me up and I sat in a chair with my coat on. He was lying on his hotel bedcovers with his head on a pillow and his hands clasped under his head, tired. One thing that should amuse you: he said that for too long he labored under the misconception that a writer of his stature had to screw everything in sight. And nearly lost his wife that way."

I take this as a rebuke, as I gradually come to see something he said earlier, "I've come to believe young men in New

York tend to employ women for entirely sanitary reasons." When it's clear this has gone past me, he adds, "To keep from dirtying their own bed linen."

Van says, "I have another favor to ask, a dangerously secret one, and I'll pay you. Can you shinny up a pole?"

A neighborhood of a half dozen streets, under the Brooklyn Bridge on the Manhattan side, is being razed, he tells me, and one of his dear friends has lived many years on one street there and written cherished words. "So attend! I want you to find a certain street sign—I'll draw a map—and then I want you to shinny up the pole and lift it out—they slide out the top, I'm told—and carry it here to me, and you've earned yourself twenty-five smackeroos. Take a newspaper along and wrap it up! Don't let anybody see it!"

"What if I get caught?"

"Say you're exercising. Nobody lives there anymore. It's a wreck, a shambles! If a policeman sees you he'll probably let you take it. Say it's for a friend."

"Don't you have scruples about this— I mean, as a person who works and writes for *Guideposts?*"

"If we don't do this it will be ruined, lost! We must be faithful stewards of all that's given us!"

The sum is so phenomenal, as much as I make some weeks, I can't resist. And it's true, as he said, not one person is on the dismantled and concrete-strewn street, so I shinny up the pole, get it, bring it back, and he pays me. Then he says, "Now I must have you to the party!"

It's a birthday party for the writer who lived in a studio overlooking the street, Joseph Caldwell, of *In Such Dark Places*, mild and deferential, with searching eyes. He sits most of the evening at a wrought-iron table on the balcony of Van's apartment, a sumptuous layout, a floor-through, and after dinner Van carries out a cake.

"I baked it myself!" he cries, then says, "No, somewhat like George, I cannot tell a lie, anyway not biggies. I got it at the bakery and our friend got you this."

He passes him the sign, wrapped, and when it comes out of its paper, a glint of moisture covers Caldwell's eyes. He has another drink, we all do, and near the end of the night he says I might consider writing for TV, since one can make good money there and sometimes still get a decent script on, advice perhaps based on Van's description of how I am an actor who wants to write, and perhaps Van has explained, too, how destitute I am.

How can I get even a fraction of the life I lead (and only in months!) in a letter to my father? I try, I write to him nearly every week, but all reality is absent.

Today I go out to our recently built shed in a looping stumble, one way of describing any attempt to negotiate the drifts, and discover the sheep are out, in the stack of bales. They've walked over a drift above the upper fence rail. I yell in the wind and they clamber back over but one, as always, goes the opposite direction, circling my shed. I take off, able to run in an area where the snow is worn away from eddies around the shed, and when I'm almost at top speed I trip and come down on my knee on a stone, huge as a torso, with a shock of pain so great my leg is numb to the hip.

The horses are excited, outside, glancing at me in curious worry, and then, swinging their rumps to the wind, checking me sidewise. The pain recedes enough so I can stand, using the shed as help, and first I want to beat the bastard sheep black and blue. But they're low on feed. Me. A streak of heat pours over my knee, down my shin, and I'm sure it's an artery. I take hold of the knee and have to hop. No bleeding I see through my coveralls, apparently another form of pain. I can't bend it but go in a hop to the corner of the shed, where the fence meets it in a walk-through, and watch the horses blow steam and throw their heads. Frost has gathered so whitely on their whiskers they look catlike. Cody hobbles over, stiff from the cold, and nuzzles my glove as if to soak up the reason for my groaning.

"My knee!" I cry, and he jerks his head away in surprise. I reach out and rub a bead of ice from the corner of an eye. He goes in a pivot through the door of the shed, his short tail like a pathetic broom, and shoulders and bumps and tucks himself in among the other horses, the prime mover of the group, for their heat.

Sept 21, 1964. No entries now for over a week. Deep in the transition state—the change Maxwell saw in me three weeks ago. Dan and his friend, Joe, left this morning after a week's visit. A comedy of errors finding them a place to stay, keeping them busy when probably all they wanted was to be left alone. Got them free passes for the Fair and they used Tony's floor to sleep on a couple nights. But they also took a hotel room ($4 a night) so *they could get at their luggage when they wanted to. Before in a locker, or carrying it from place to place.*

Joe seems too naive to be a doctor, an innocent, or maybe it is the city's effect. One evening, trying to provide entertainment, I call Bob, who has moved to an apartment on 14th, a plain brick building unlike the architectural specimen where he once lived but with much more room, and ask if I can bring my brother and a friend up for some wine.

"Great!"

His place is still only partly furnished, with the antique circular oak table main-room center, and all four of us sit at it with a jug of wine. Bob gives me a high sign, one of the faintly dramatic facial expressions we use to communicate, suggesting we attempt an act we practiced one night: levitation of the table.

"Hey, let's try something," he says. "But you guys, you really got to concentrate." This is to Dan and Joe: I know the drill. "I don't know, though," he says, and shakes his head, looking at me with a glint. "How do you feel?"

"Focused on the inner."

"Here's what we have to do," he says. "You all put your hands on the table like this and really I mean *really* concentrate." He places his fingertips in a spidery stance and we copy him. Then he draws his chair in closer and I know he's getting his knees in place under the supports that angle out from the pedestal leg. I do the same on my side, as we've practiced. We close our eyes and moan and grunt—the weight is considerable—and a wobbly trembling passes through the table as we balance our lift of it against each other, then we get it off the floor, whoosh, my feet trembling on tiptoes.

Joe shoves back his chair and jumps up, eyes bugging, and cries, "Cripes, guys! It's *doing* it!"

"Well," my brother says, "I think maybe you should pull a Houdini, you know, and check underneath."

The finest time is Saturday, with Dan in my room on St. Marks. After attempting to define our unity, we start reconstructing the past, drawing sketches of our house in Sykeston and correcting each other; putting together two sides to every event, calling up people's names, in the looping but always connected momentum that comes when two who have lived together for decades enter their past.

Outside the windows, sheer black: crisp cold fall night. We reconstruct enough so I know that my memory's not awry. He returns to his hotel and must miss making connections when he and Joe leave in the morning—perhaps he knocks and can't raise me: my recent overwhelming dreams. So in my reconstruction of the hour, I see him purse his lips with disappointment, then slide my New York map under the door.

There it lies on the linoleum when I wake.

Gone, I think. *How can I possibly tell all this to Dad?*

A Definition of Internal Bleeding

Sept 24, 2 A.M. Tuesday, yesterday, Maxwell received my letter and called right away, asking me to come either for lunch or at 4:00. I don't know why I chose four; I hadn't eaten and didn't have any money. He wanted to see me because of a paragraph in my letter—

"I think I've ruined most of my fiction. Instead of being honest I've been trying to be profound. It's impossible to be profound if you're not. Besides, it seems as soon as you try to be profound about people, since they are what writing is about, you miss all their humanity."

He was overjoyed to hear "the good news" . . .

He wants me to write a factual account of the last week, when Dan was here, and I've been cursing myself for not keeping this notebook during those days. He wants me to write simple exercises, "Just for me, without any thought of publication."

All we have left for hay after using our handy square bales is a stack of the huge rolled-up affairs that look like monstrous shredded wheat, and they're so far from the shed we finished through ice as winter hit, I don't see how we can get them to the animals or the animals to them. But Joseph backs the Ford tractor from the north garage where he parks it and starts the blower, sending a twenty-foot arc of white up into a shimmering fan, the rainbow tints at its edges appearing frozen in the cold. He finds a way around the fences past the garage, then hooks a series of chains to a round bale and drags it through an acre-long channel he has blown to the barnyard, a bale a day. But each day he has to blow his way out and blow out the route to drag the bales along.

Maxwell and I call the pages he believes could be my first novel "My Brother's Visit to New York." I look at *New Yorkers* on newsstands or buy issues when I can, and it seems a story I wrote in a rush for Scouffas, "Requiem and Fall," is closest to what they do. I prune out the worst literary passages, which I'm learning to identify, retype it, and address it to Mr. William Maxwell, The New Yorker, etc.

The next day I sit at Attention under the hypnosis of collating pages when Austra says I have a call: Maxwell.

"Can you come up?" he asks. *They're going to take it*, is my first thought. "We'll look at what I've done to your story and go for lunch. Would sandwiches in Central Park be all right? We'll make it a picnic."

He has gone over the story sentence by sentence, and I'm able to understand what he's after once we look at three or four pages, and he senses that. "Can you work on this in addition to 'My Brother's Visit to New York'?"

"Sure, I think."

We sit on a bench in Central Park and he pulls wrapped sandwiches from the paper bag in his lap. I undo mine but can hardly eat. I can't talk; an iron hand lies on my chest and throat. I feel I have to cough but if I cough I'll weep. All he has done for me presses like the hand of iron and now the time he takes to sit with me, feed me, well. A pair of swans is circling on the lagoon below as they did on the cover of a novel by Elizabeth Bowen he earlier bought for me and I feel choked worse than on the day I tried to ask for a job. The swans in their orbits are my focus and I hear his whispery voice and picture his detailed comments on my pages in his rapid hand and see Bowen's title appear over the swans, *The Death of the Heart*. That is what is happening over this time that feels so endless I seem to remain on the bench. I turn and his forehead is set with its wrinkles of concern, the V at dead center, and he pauses, his lips parted, as sometimes happens when he searches for a word, and I can't take in what he's saying, as if pinned to another time.

("When we used to sit on a bench in Central Park," he wrote in a recent letter, "and talk about writing I felt I was engaged in a kind of Faustian struggle with your soul, with over-intellectuality exerting a strong pull on you.")

It is not death but a beginning stirring of life so potent I feel I'll gag on it and then feel drawn from that by his words falling over me with the force of love. How can I leave? What would I do? Where would I go? We are on a path, then in a cab, and I see how he tips, studying the meter and counting out the exact amount, his concern for the cabby of a part with his concern for me, then I remember seeing my pages on his table, then leaving, breath held.

And by that magic transport that can happen in a car, when you're fifty miles farther down the road with not a shred of memory of how you got there, I'm in my room, staring at the pages on my bedside table. Every suggestion of his, every interleaved word in the typed lines turn me toward a point I know I have to reach. "Sometimes it feels, I know, like internal bleeding," I hear him say, and look up. It's his voice and I'm not sure if he said it today or another time.

But I hear it now. I follow his trail of pencil marks, feeling chunks of words and paragraphs break free and fall from the ceiling of my mind, and when I look up, it's dark outside. My windows show me back in shining black plates. I lie down on the bed in all my clothes. *Care.*

I wake with the chilly feeling of a child awakened the hour before sunrise for a vacation or a fishing trip, when whatever mechanism it is that keeps heat from leaking from your bones hasn't kicked in yet. I take my closest subway, the Lexington line, to 86th and get out in Yorkville, Germantown, his neighborhood—he is home for the rest of the week, I know—and as I walk, *oh, that towering feeling*, turns in my head, *that* o-ver-*power*-ing *feeling! Knowing I'm! on the street where you live!*

I walk into Gracie Square, a public park, and notice that the fur on the tail of a squirrel is so fine it seems transparent; I see through it to the ground, to the veins of leaves lying in mud, the tail a mirage.

I walk to the river and sit on a green-slatted bench like the one in the park. The sullen, slow river, myself and my senses coming to themselves in fall air, as they will every year after, in commemoration of this day. I will remember every single hour with him, I think. *Mark that down.* I will return to others, as much as I am able, something of the love he has lavished on me.

Then I walk back to Lex and 86th and as I pass a German grill I feel famished. Two dollars in my pocket. I go in and order the only item in my range on the placard overhead and get an odd look from an abattoir-master at the counter. He sets down raw hamburger mixed with minced onion on a slab of white bread, an open sandwich, one of those awful German delicacies I abhor.

Then the surly meister stands there watching, so I eat it, every bit of it, my first and last of ethnicity, and in my hunger no meal tastes quite so good. Out in the street, giddy and dazed, as if I've downed all the beers lined on the counter in front of the elderly men, but no, I haven't, not a drop. I feel again the pain of the iron hand, now directly over my heart, and know it's not the death of it but the opposite.

I haven't seen Dylano in a while and feel the sway of her as I've seen currents off the beach take hold of and sway swollen pods of seaweed, the weeds themselves barely visible below the blue-aqua sea branched with traceries of foam, then pull them under with splashing pops like gunshots across a wave. So I take the subway out and psychic contacts start to spark. At her door I hear voices and when she opens up, her face set differently, I'm staring at a monkey on her back, its head above a shoulder, pawing her orange ringlets, its dark eyes the size of half dollars.

She says, "I want you to meet two new friends. This is Oscar, my kinkajou"—she turns—"and this is Billy."

A guy who looks sixteen walks from the shadow of the kitchen with the bowed melon grin of the gratified.

Sept. So many major conflicts are being resolved in my dreams, and my dreams are filled with Maxwell.

Remember these: the one in which he is killed in the barren house, in the north, by savages. His friends deserted him for some reason. Somehow I knew I could have saved him, and somehow felt responsible for his death; I knew the customs of the savages and he didn't.

Then last night the dream with the razor. Maxwell, disguised as a woman, I think, cut me with a razor. I tried to get it away from him and he kept cutting me. It was impossible to get it out of his hands. I crushed his fingers and he said, "Don't," because this gave him the opportunity to slash me more. And because the wounds he was getting didn't hurt him and mine were detrimental.

I wanted to get the razor away. That's all. . . . no malice intended. If I got it I was only going to throw it out the window (this seemed to take place at Tony's). The cuts—on my wrists, across my chest, stomach, and shoulders, even on my face, I think—finally got so bad I awakened. In a somnambulist state I still was thinking of ways to get the razor from his hand without hurting him.

No more promises to myself, no more commands or aphorisms to anyone including myself. No more sanctimonious judgments. No more magnanimous advice.

But look at what I've just written.

"I'm writing something, too," Bob says, sitting down at a small table in the side room he uses both as bedroom and storage for his costumes and props, and I figure I must have interrupted him when I buzzed. "It's a kind of journal."

"Of what you're doing?"

"No, like a novel. The journal is a journal this weird guy keeps." He gives his mixed smiling expression of pain.

What a person is in an image is not what a person is in words, and neither is the person in the flesh. The word was made flesh only once but has filled every writer since with the belief that words will represent a person or series of persons as characters *in the flesh.*

3:30 AM, *Oct 4, 1964. Forty-six pages done up to this point. A flood of ideas. Creating a myth from "true facts."*

Have been very sick. I fear the probability of gonorrhea. Never in my life had VD. One of those cruel unreasonable doctors who make you ashamed of being sick prescribed an antibiotic and complete rest. That second part is impossible. I'm feeling somewhat better today. Also, he warned, no drink or sex. Neither of those are difficult. No money, no woman.

Doctor five dollars ("Do you make much money?" "No.") and prescription six.

St Marks, Oct 5, 1964. Slept for nearly twenty-four hours. Must be sick after all. Fifty-five pages about My Brother.

. . .

My face is heated with a fever like shame, as though the truth that bodies become one is infinitely true, mine half rotting, rot. I walk the aisles of a deli, unloading the shelves with my eyes, buy a nickel candy bar and dash out, and in my room press my face into my work, my bed of rest.

St Marks, Oct 11. And I, the thundering bridegroom of your empty bed, rage and rage against you, and grow tired, and wish only for sleep.

Maxwell sent me to his doctor on 60th St, Jack Nelson. An impeccable man, of course. I had crabs, a gonococcus—most of which was gone due to the chloromycetin from the other doctor—and a bad cold. In order to destroy every possible ramification, he's giving me the two week penicillin treatment for syphilis . . . A lesson.

The limitlessness of Maxwell's compassion. Also loaned me $15, pulled from his billfold, and is paying for Nelson until I have the dough. I must have looked in sad shape because he stared at me and tears came to his eyes. He only wanted to talk to me, to see how the writing has been going.

75 pages on My Brother

12 pages on the new "Requiem"

Bob sits across from me at a table draped with a heavy white cloth long enough to cover our knees, in the back room of a restaurant of the kind neither of us could afford if it weren't for a bounty from my Grandma Johnston's estate, and repeats again the details of his unhappy parting with the person he calls "my French girl," drawing me by the gravity of his magnetism down dark alleyways of my unsatisfactory or possessive turns with women.

As he talks he takes bread from the covered basket between us and breaks and tears and shreds it over his heavy dinner plate, never placing a bite in his mouth, pulling, twisting, fragmenting as he stares in sad fury at me.

Our wine comes and when I go to pour he sees his plate, heaped with shreds and crumbs. "Oh," he says, and raises his hands like a priest giving a blessing, but as if to hold the plate at a distance. "I did it again."

2

I write more daily pages than ever, all background, prelude to my brother's visit, and then a phrase sends a scene wobbling up in focus: my grandmother, my mother's mother, at a sink in her farmhouse. Then she appears in sunlit brilliance and the scene runs through, an exchange with a grandson at her side, and I know I have a story. I will write about her the way I've always felt I should, as if my life depends on it. Enough of fractured time and the grotesque and multiple viewpoints and trying what Beckett and Babel and Kafka have done as only they can do, besides Nabokov—just get down clear the slant of light on a woman who influenced me more than any writer.

And if I can't admit that and record her with as much accuracy as I can summon, then her life and a host of others connected to it will be lost, gone without a trace, as a last lesson is wiped from a blackboard with a wet cloth.

I write the story in one sitting, type it up and mail it to Maxwell, adding as with Ebert handwritten changes, now on the gold-colored paper I use for drafting—for no reason than that it's free at Attention; white means a final draft, my method to keep matters sorted, and as I continue my walk to 43rd I feel the imagined excitement of Maxwell telling me I can move to a final version, that startling white. A mist fills the air and bears me up in a way I felt only as a child after communion or when I walked in the woods in the communion of not consuming a host.

I enter the arch of the building from the south or 43rd Street side and take the elevator, the one operated by a black man for Mr. Shawn, who mistrusts self-run elevators with their buttons. Anybody who rides up or down in this one maintains the silence of cathedral worship, staring up at the dinging dial as at an icon—absolute silence for all the years I rode in it, except for once, when an elderly man with slant shoulders and gray-white eyes jabbered in Spanish to a companion all the way up, and when the doors opened an editor was waiting, and said, "Ah, Mr. Borges!"

I give my name to the secretary behind her window with a half-moon bottom like a subway-token booth, and Julia, with her dazed look of far-sightedness appears and leads me down a green-painted hall floored with brown-maroon linoleum to Maxwell's office. Around his desk in one of those leaps of movement that astonish me, he shakes my hand, and is behind the desk by the time I've sunk into the wooden chair, my wool cap in my lap.

"Well," he said. "You're in."

Because I've worked as I have I'm not sure I've heard him correctly. "What do you mean?"

"In. In *The New Yorker.*"

"No."

"They liked it a lot."

"Who's they? I don't understand the structure here." My first reaction, to hold it at arm's length. "When you wrote 'I'm sending it along' it sounded ominous."

"Two editors read it, generally, and write what they think of it. Then it goes to Shawn for his decision—based on what the others have said."

"He approves everything that goes in?"

"He gives his pass on it, yes." He turns somber, a line of straight wrinkles ridging his forehead, with the perfect V at their center. "I thought the story was awfully good but I didn't want to tell you that and get your hopes up. It broke my heart every time your poems came back. I liked them but about poetry I can do nothing."

"I don't know if I should tell anyone." As now the news arrives. "Maybe I'll wait till it's published."

"It's up to you. I've only told one person. Do you know who?"

"*Chuck Shattuck,*" we both say at the same moment.

"That's what I thought," I say as he says, "Yes."

He says, "I stole that pleasure from you. As soon as I heard I sent him a letter."

"I hope it makes him happy."

"It will."

I realize I have to write him, too, to smooth any wrinkles between us, though none seem to remain.

He pulls the story out, I catch a glimpse of a curl of my handwriting, and the weight of what he's saying hits. There the story is, with its hue of gold-orange tulips.

"The pay doesn't sound so good but it will mount up. You get paid retroactively with your next acceptance. I'm having a check made right now for three hundred dollars. Usually the writer has to wait a bit but I thought you could use the money. You can pick it up tomorrow. By the time everything is over, you'll be paid a bit over five hundred for the story this time around, and more later, then more with the fourth, the sixth, and so on. They pay by the word and I got you the highest rate without a first-reading agreement. I have some say in that."

"Thank you."

"You deserve it. And I'm having a first-reading agreement made out. That means, besides the word rate, you'll get a cost of living adjustment and so on."

"What can I say?"

"Nothing. Just don't let your feet touch the ground."

"What do you mean?"

"Float."

I start to stand and sink. "I can't think of anything profound."

He laughs, then starts around his desk and with the movement his windows revolve. I'm up and take his hand. "Thank you, if it weren't for you, this never—"

Emotion tackles me and I'm in the chair, my face in the fur cap as if it's my grandmother, my father, also a mother to me, I understand now. In spite of our misunderstandings, he never wrote me off but prayed for me as a mother does. "I'm sorry, I didn't think that would happen."

"I've been trying to find a Kleenex but there isn't a one anywhere."

"I used this." I hold up the cap.

"I mean for me."

I hold up the cap. "Do you like it?"

"Lovely."

"I bought it with money from my grandmother's estate, the one in the story. This coat, too. Do you like it?"

"Yes. You enjoy nice clothes, don't you?"

"I never had them. That's not true. My mother had excellent taste and my grandmother, too, the one who paid for these, and she gave clothes for my birthday, for Christmas—after my mother died, the best gifts."

"I've enjoyed watching the way you dress. Your taste is aristocratic. And don't be apologetic for your feelings. You aren't a writer if you can't feel as you do. When you stop feeling, you stop writing."

It was near the end of the working day when his phone call came and now the room is the violet of late November. He hasn't turned on the lights and looks dim hardly an arm's reach away. Twenty stories below is the street, not a sound, and as the violet invades us further we fall into a conversation difficult to record, where nuances are understood and only a word here or there keeps it moving.

"I like the ending."

"Good."

"They should blow fuses."

"?"

"Like the last words of a poem."

"I see."

"You have the gift."

"Oh?"

"It's in you. As in the story. Your endings do it."

I rise to go, to free him to leave, and he says, "It was three years ago Chuck wrote to me about you."

"That long?"

"I think so. It was the day he took you to the Co-op Bookstore and bought my books for you."

"The writing contest, when I was a sophomore, 61, yes, that was three years ago."

"He wrote, 'I have a writer for you.'"

"I was afraid it wouldn't happen."

He shakes his head, sadly it seems.

"You said things that startled me, because they assumed I was the person I wanted to be and when I left, I thought, By God, he's right! So I was able to keep going."

"Nobody can do it for you. It's been difficult, I know, but here you are."

"In."

"Yes."

"Maybe there is a divinity that shapes our ends."

"I suspect so."

"When I think of what . . . "

"Your mother's death affected you more than you can admit yet, I believe. It's partly the reason you write. It was only a matter of time before you were published, I can assure you of that."

"When I think of—" What I was writing only a few months ago, I want to say, but see he understands, by his grave nod. Again I say, "Thanks."

He nods in acknowledgment in the darkness that has all but enveloped us.

Then what name do I attach to my work? "Larry" is too diminutive, for my taste, having lived with it, and "L. A." has the tinge of a creep toward Eliot and "L. Alfred" a pompous king. I remember Tyner always called me "L", and Maxwell signs his letters "B," and then I walk into a Slavic bookstore in my neighborhood and see, lying on a table inside the door, *The Cossacks*, by L. Tolstoi, and that does it. I tell Maxwell.

"It doesn't much matter what you put at the beginning," he says. "It's Woiwode they will remember."

I'm at a table at Attention, collating and stapling with the giddy sense that I might decide not to continue, when Van comes swaggering up in his twitching way, whistling. He leans to my ear. "You're the greatest!"

"Is that what you were trying to remember?" All day he's been saying there was something he wanted to tell me but couldn't remember.

"Not at all! That's not it at all! No, I believe in giving credit where credit's due!"

"Oh, thanks!" I've fallen into the game he seems to be playing, or anyway the rhythm of it. "Thank you very much!"

"I don't wait till people are dead."

It hits like lightning. He wiggles his orange eyebrows and goes waltzing off. I'm sure he means the story about my grandmother. Such statements, when they came then, tended to undo me, but now seem the working of the Spirit, for those attuned to it, since the Spirit is infinite, beyond what we assign by rigid or free-floating theologies. It was a message necessary for me to hear: *Give credit where credit is due, to the living; don't wait till they're dead*. Love your neighbor at least as much as the skin that holds you here while it holds evils, often microscopic, the worst ones, ones you're unaware of, out.

I want a prose trim enough to follow the swiftest physical action with ease, yet with enough substance to pull a passage into a dimension a character assumes (with a resonance to suggest the range of his or her feeling), and the ability, now and then, to spring into reaches of far-flung thought. These register what it means for the physical world to have a sky: transcendence, abode of greater beings.

Purge falsity and rhetoric from it, whatever the cost, even if such raw emotion remains that other men cry, *Sentimental!* Keep to the simple surface texture of Gide, the compact phrasing of Colette, pure poetry in the guise of speech, never using any word that doesn't fall straight from your brain to your tongue.

An impossible task. I worry myself less with it then than later, because it's always time to finish up, get done, send it on, get a life. But I keep at it on and off until the day I hear the auditory echo exerting pressure on my eardrums from the inside as camshafts of phrases turn within the whole of a sentence revolving along its length to a point where no iron can so pierce the heart (to paraphrase Babel) as a period put in the right place. I want to see this on every page, in every paragraph, in sentences, at least in their balancing acts against themselves. The gait of thought, the pace of pondering, all the recreational rest of real writing.

I'm reading Renard, jaunty at my first sale, and write a story in a week. It has the feel of that. Maxwell gently says it will take work. I start another, chastened, about a night my mother is taken to a hospital, trying to listen to each sentence. Next to the story I have a sheet of paper handy because the title keeps trying to pop through and I try to catch it, and toward the end of the draft I spend as much time listening to it sink as to the draft. Then I put down the last sentence and it rumbles out as if lugging logs behind on tractor chains, *Beyond the Bedroom Wall.*

Dear Larry,

11–24–64

What a night and what a day!

Congratulations again. At a time like that a telephone is a wonderful thing and yet how inadequate. I'm sorry I can't transmit my *real* feelings better but I guess everyone around me today knew how I felt. I was either walking on air or crying for joy. What more can one do?

Thanks so much for thinking of me because believe me there has never been a day passed since you left that I haven't thought of you. Mostly wishing and hoping that no matter what happened or what others may have thought you'd know my confidence in you never wavered. I only hoped that your strength would hold out.

It's from my father, four more handwritten pages I want to savor so I go to the window of my room and stare out as if I'll see him as I saw him at the window of our house with my first poem fluttering down, a fledgling in his hand.

Maxwell says, "When we were in France, in Paris, I went to the Luxembourg Gardens and stared up at the balcony where I knew Colette had an apartment—that ageless practice."

"What's that?"

"You always stare at the window of the one you love."

I ask Bob about a watercolor on the front wall of his new apartment—twelve by eighteen inches—of a man seated (the pale mat within the thick black frame severing his upper forehead) in a tan cape that falls over his lap and down through the rest of the portrait in an uncoiling of unassailable lines, the colors muted, as if the man is singing a soft song in the blue and green space he's given to occupy. "It's my dad's," Bob says.

"I love it. What does he call it?"

"*The Actor*."

"It looks like you."

"More like you. No. I mean it. Kingly."

Of course I've described in detail all my roles.

The Maxwells have me to their place on Gracie Square for dinner, and I meet his wife Emmy, a raven beauty, and

Katharine, ten and the image of Maxwell, and Brooke, eight and the image of Em. After dinner we talk about MacDowell Colony, and a fellowship he hopes to get for me to attend, and Kate says, "It seems to me that once a writer got all by himself he'd find out he didn't want to write."

As the Maxwells put their daughters to bed I fly across the room and pull down a book that's been winking from a shelf, an Updike, and find on the title page, "Dear Bill, New York for me will always be the New York out the window of your apartment on Gracie Square."

3

"Let's go get the inverter from Jim," I say to Joseph. Jim of the Air Force has one and I'm feeling brave with Joseph and Ruth home, after our successful run for wood.

"Dad, you don't know how bad those roads are. I do."

"Hey, the mother cat's come back so this can't last." This is true, the mother of the three kittens is back in the doghouse, but I put her there, after hearing something under a snowbank and digging down to hit the furry hide of a deer, a road kill I forget I'd lugged back for the dog, and there she was, inside a cavity between ribs that she'd eaten hollow, glaring at me over her shoulder as if she wanted to burp in peace.

So we head down the road and in a mile have to roll down the windows to search out the ditch, our only guide, which I start into once. By the time we get to the bridge over the Cannonball, banked at the far side up to its rail, we have to turn around, but have no room except on the bridge, then get stuck, no shovel along and the cold so bitter anyway you couldn't dig, and only Joseph's mighty pushing frees us, and when we head back in a wind that's so fierce it feels my hair is streaming back from its force through the windshield, I know we couldn't have made it more than a mile walking in this, and then in the house I see how pale Joseph is, shaken with a fear I've never seen in him, and he says in a voice trembling from that or anger at me, "I'm never going to do that again."

I'm introspective, reticent by temperament and type of work, with a skepticism about human nature, including my own, that can lead to depression. But during this time of concentrated writing, when my material arrived in such great chunks that sentences and paragraphs fell full-blown from my sensibility, I felt radiant, giving off a glow as I ran up the five flights to my room at St. Marks or rode the elevator twenty floors to Maxwell's office.

It's given to a writer only two or three times in a life to breathe the fresh air of a newfound field of material, and this was my first flush of that wholehearted yet disciplined freedom—no new land of air to breathe without continued work. And that is the way it went in the first years of my relationship with Maxwell at *The New Yorker*. I called him Bill in his office and in letters, because he asked me to, Maxwell to my wife and others, to keep him to myself, and to honor his office as mentor.

He had been at the magazine for thirty years and for most of those years was in only Monday to Wednesday and at his own work the rest of the week. His books are cherished by devoted readers and Updike has said "his voice, simultaneously very personal and clairvoyant, is one of the wisest in American fiction. As well as one of the kindest."

He worked with Updike, the Franks O'Connor and O'Hara, Welty, Salinger, Cheever, Brodkey, Sylvia Townsend Warner, and several dozen others, and had as much influence on the direction of American fiction from the forties into this century as any single person, with no trumpeting or manifestos but his standards and clarity of thought, and came to be the senior fiction editor—if such a designation can be given to anyone in *The New Yorker*'s formerly loose structure, with everything feeding to the one at the top, first Harold Ross and then William Shawn.

I rode up the elevator radiant, feeling my face gave off light, and the receptionists, by now familiar with me, rang up Maxwell and I was sent through and walked down dim "Sleepy Hollow," the hall to his office, and as I came through the door he was around his desk (a wide oak one from schooldays) in one motion, with the springy grace of a person in the first years of achievement, and had my hand in his.

Then the work. When a story was taken, he had it set as it stood, with typos and gaffes removed, in rough proof—so designated to the typesetter. My manuscripts were fairly clean, with a few handwritten interlineations, and I can spell. The idea that writers are terrible spellers is another of the Hemingway-Fitzgerald myths, and it's only in recent years, past fifty-five, that I've had to pause to wonder which letter within what word goes where. Plus those damn apostrophes.

He would mail me a copy stamped "rough proof" at the top with suggestions written to the sides of the column of type, as it would appear in the magazine, set down the center of a typing-paper page, or place bracketed questions within the text itself. Rough proofs were for my indulgence, mostly. A new sentence would arrive, a conversation open up, and sometimes new paragraphs or whole pages fall in place.

Using words to maintain one's slant on the metaphor a story takes is a process always in flux. A story continues to reveal itself and to grow as a child does, even after the child is an adult, and I've never published a story or book I didn't want to fix once it was in print. So the process of working through a series of galleys wasn't only indulgence but a way to work out the worst of the wrinkles that were such irritants once a piece was published.

I would finish and mail the proofs in or, more often, carry them up myself, usually in the afternoon, after 2:00. At 1:00 Maxwell removed his shoes, donned an eye-blind, and napped for an hour on his cream silk couch. I tended to arrive unannounced and often interrupted other work, I'm sure. Once he was escorting a handsome woman down the hall and paused to introduce Mrs. Francis Steegmuller and it wasn't until we were in his office that I realized she was one of the regular writers I most admired, Shirley Hazzard.

Another time he said somebody was stopping by and a gray-haired woman with the portly build of Colette in her grande dame phase, came in an imperious way right up to my face and cocked her head and stared into me as Mr. Shawn once did, and said, "I had to see for myself. Thank you."

It was Janet Flanner, Maxwell said—the provider of "Letter from Paris" under the pseudonym of Genêt. They were the only writers I met (other than the Borgesian encounter with Borges) on my many visits to the office, though I saw Joseph Mitchell and Philip Hamburger and Brendan Gill and other regulars on the staff at social events.

Maxwell's voice was modulated to the intimacy of office conversation, with its breathiness and the hesitations that had at times the appearance of a stammer, and a relaxed and open glottal caress at its base that made something gently exceptional of language, as when he said "marvelous," so that it sounded exactly like what it meant. And his eyes, below uncurved, straight thick eyebrows—large and edged with lashes like kohl, deep brown, brimming with empathy, the sort that drew you in to discover its limits.

I never did. And I sometimes sensed he could be easily taken in, or taken advantage of, although the evidence was against this, considering those he handled, including, in his American incarnation, Nabokov. I took an upright wooden chair to one side of his desk, near the wall with windows looking down onto 44th, and talked for a while, about writing or family, and this no doubt gave him a chance to assess my mood and state of mind.

A decade later I mentioned that when I walked into his office he sometimes said, after that sweep around his desk, "Pardon me a moment," and returned to his chair, a creaky swivel one of oak, and swung to his metal typewriter stand, where an unfinished letter was waiting, and finished it, and he said, "Was I so rude that I wrote a letter with you sitting there?" I remembered he had a number a times, yes, I said, but I never felt it was rude; it suggested his ease with me, I said. But it was then I realized how often I must have interrupted him walking in off the street.

He finished letters because of his diligence in keeping in touch with his writers; any letter he received he answered that day. Or a letter from him would arrive in the mail, unexpectedly, about a book or a passage in a novel it occurred to him I might read, to help me along on a story that so far hadn't worked, or with an idea for a story I hadn't yet discovered, or simply news.

To type he slipped on his reading glasses and paused a moment, regathering his thought, his fine fingers pointing down at the keys rather than resting on them, and then rapped out sentences with such rapidity the upright portable set the rickety metal typewriter stand rattling. Then he rolled out the sheet, usually a half sheet of *New Yorker* stationery, pulled out his pen, and signed it *W.M.* or *Bill* or merely *B.*

On one occasion when I had a drink too many, which could mean two, since I had the volatile blood sugar of diabetics from my mother's side, I said in a stumblebum's voice, "I feel like you're my literary father. I have my regular father, naturally, you know, but you're my literary father. But, you know, you're Updike's father, too."

I was being imprudent, I knew, but he merely said, "Oh, no, it's quite clear John loves his father to beat the band. Which is fine by me."

It wasn't for years, after my own father died, that my untoward confession rose and I heard his reply as a rebuke: that I did not love my father to beat the band.

I was stepping beyond boundaries, breaking an unwritten law, because in his office, as in his letters, no gossip or talk about other writers he worked with went on. Each was kept separate, his or her life sacrosanct. He seldom went to parties or literary soirées: "I like to get to bed early, and one drink puts me to sleep." He had a time for tea (which I was invited to join in, when present) and a time for his nap, and the time each week for his own work.

As he talked he would lean back in his chair, with his hands clasped behind his head, and a subject that often came up was his family, or yours. Which led me later to the understanding, without his saying it, that any committed writer who also has a family has to make decisions that take on the weight of earth's gravity about the balance of a book or a career, say, against the illness of a child or merely rearing the child responsibly; and the central relationship to one's spouse. Then the more tenuous but weighty one of faith and career, and if your career is your faith, then your God is as good as dead, once your work is, and so it goes.

Often the phone went off in jerky rings. It was one of those old-fashioned black chunks, pushed to one side of his desk, and he reached for its receiver with both hands as for something distasteful or anyway untrustworthy he wanted to be sure he had a good hold on, and lifted it with both hands to an ear, gingerly, switching to one hand only when he had to reach for something on his desk.

Once he said, "Yes, I have them right here," and pulled out a set of pink galleys. Pink meant a story was edited and past the checkers and ready to go. He flipped through pages, making additions here and there, saying, "Yes. Yes, fine, I think." And then, "No, I believe 'womanly' is exactly what you intended. I'd leave it." More changes came through the receiver—not unusual for a writer to call in even after final page proofs—and then he set the receiver down, again with both hands, and folded over the top sheet of galleys and I was able to read, printed lengthwise along the column of type, as longer titles were, "The Bulgarian Poetess."

A quintessential Updike story, as I learned when it appeared in a few weeks with "womanly" intact. The magazine came now in a complimentary subscription, one of the perks of a contract writer. It was a true weekly, then, and two pieces of fiction appeared up front, one short and one longer (with sometimes an additional story near the back) which meant at least one hundred stories a year. Maxwell mentioned this whenever I was disaffected with one, as if to say out of a hundred they can't all be what everybody wants.

Most editing went as easily as what I saw happen over the phone, but in the years I was in the city I walked up, except for occasional fixes on page proofs (these were the actual pages laid out, including the "art" or cartoons), when we were under time constraints. My phone would ring and that whispery voice with its glottal caress would go into the matter at hand, often without the formality of a hello. Then he hung up when done, *click*, and you were holding the receiver or saying goodbye down empty wire.

The editing in his office took place at the oak table set against the wall inside the door. His desk faced it, so when the preliminary talk was done, I would rise from my chair and turn it to the table and he would pull up an identical one and sit beside me, placing me at his right hand. Only a pair of galleys and a can of pencils sat on the table, and there was no aura of august or lingering literary presences. All was swept clean by being irreproachably Maxwell's, or anyway he made you feel the only important matter was your story on his table.

A wooden hat rack to the left, at the head of the couch, held his hat and overcoat—or raincoat and umbrella, depending on the weather. Another battered black umbrella leaned in the corner, sprung open and wrinkled, ribs bent, available if you needed it, sometimes with clinging drops sliding over it after a rainy lunch. A rectangular clock with a touring car, a driver in a duster, and hoop-skirted passengers printed on it—a reproduction of a turn-of-the-century ad with "The 1908 Maxwell" across its base—traveled to different areas of the office without ever attaching itself to a wall, to my knowledge—a gift from a writer. 1908 is the year of Maxwell's birth.

A set of the galleys on the table were his, another mine, and when a decision was reached between us the words were transferred to the master. That was his. His way of working was to gently bring you to understand how you might have done it if you had the power and perseverance (and fewer of those romantic entanglements) of the masters he admired. "Oh, you haven't read Conrad's *Victory*? Well you must, and learn how he deals with women of this stature. Read it, and this paragraph will change." As it did.

"Think of how Colette would have phrased this."

Entering a character on a page is similar to an actor's trick of moving in and out of a role without upsetting the sovereignty of self, and with my own roles in words I wasn't pulled into a foreign otherworld as when I gave myself up to a part in fire and smoke. Writers may be more vain than actors (perhaps), at least about their early work, and this causes blindnesses only another can detect. Besides, in any beginning work it's difficult to keep from encountering hostility—that raw unfamiliar interior others encounter—much less a neutral stance, and he was not only neutral but an encourager, ready to help. His way of working was to leave me with the impression that this is what I would have thought of sooner or later, pointing me to a higher place.

He could enter what was in front of him, no matter its mode or style, with his critical faculties alert and bring it to its individual pitch without intruding on its integrity, a gift! And impossible to explain, though now I can say I understand I bore inexpressible knots in myself, so deep and unreachable and sometimes malevolent that only words could release them, and he helped me find my way to the words. Set down right, they had the transfiguring power of love.

The sear of malevolence may produce a masterpiece but any continuing work has its roots in charity. "No matter how awful this character is, you have to love him," he would say. Or "No character is complete without the writer's unconditional love."

Sometimes he read a passage and turned to me with moisture beading the fringe of an eyelash, which he knocked away with a curled forefinger, then said, "I don't see how you can improve on that." He said he suffered as an editor from "guardian angelism," because he sensed he was put on earth to protect writers from ruining their best work by afterthoughts.

So at a point revisions stopped.

If he found what he felt were questionable queries from a checker, he took my side—"These people are wonderful grammarians but don't know the first thing about fiction"—and then didn't, if a sentence wasn't clear or accurate. "Isn't there a simpler way of saying it?" he would ask. Or "Isn't there a simpler way to say that?" At times there wasn't, but often as not there was. And often, too, when I would tell him what I thought I had said, he jotted it down on his set of galleys, and noted, "That's so much clearer!"

"This simply won't do," he would say. Or "People don't talk like this." Or "You're going to have to redo this entire paragraph so it makes you get out of your chair and pace around after each sentence, because you've got every one just right."

What a writer often needs, and especially a beginner, is an answer to technical difficulties. These came so often an area of my mind is made up of "Maxwell's Maxims," as I call them: "The only way to write is by the seat of your pants." This was for when I was theoretical or got explanatory. Or "A real writer almost invariably knows where to start a story. Trust your instincts on that." Or "A short story should be written in one sitting, then spend no more than a month fixing it up."

"When you try to explain away one mystery, you only make room for another."

It was an aid in finding the placement of my voice to write letters to him and, when I wasn't doing that, to carry on an internal dialogue—the massed maxims responding to a page of prose, once done, as it lay in front of me, and then to every sentence in it. As I worked on a final draft, if I installed in my mind a sense of him beside me at his table, looking over my shoulder as I worked, word after word peeled away from the essence of what I hoped to catch.

One sunny afternoon, he said, "If you're writing a novel, let me know, and I'll get out of your way."

"You mean you don't want to read it." He once said he did not comment on the novels of his writers—"To keep myself out of hot water."

"Anybody or anything that gets in front of a writer working on a novel ends up in it. Novels are all-consuming. Warn me. I'll move."

Once our decisions were made, they were final, and not the placement of a comma was changed without my approval. So my relationship with the magazine was congenial. It paid a living wage. It seemed to please him to mail off a check, always accompanied by a note. The word rate was a dollar, roughly, depending on how many pieces you sold, plus a COLA, or cost-of-living-adjustment, based on the rate of inflation and usually issued in quarterly payments. Plus "bonus" payments that added a percentage to your total sales for a year to each story over four, then six, then twelve—a process so lucrative to even the mildly prolific he referred to it as "the slot machine."

Plus an added bonus for signing a yearly contract, which stated, in essence, that you would accept all payments and bonuses. And other perks, such as the black-and-green mottled Venus drawing pencils, perfectly leaded for writing, of a grade and hardness you can't find on the street—a luxurious instrument I was encouraged by him to pocket.

And lunches in the Century Club, down the street and, if you were a good old boy and lusted after it, perhaps a nomination to the Club. It's no wonder Nabokov said in another context, speaking from his experience on several continents, that *The New Yorker* was "the kindest magazine in the world." The kindness was best expressed, to fiction writers of every stripe, by Maxwell who, as much as that's possible in a human being, was kindness in the flesh.

I saw him angry only once. When I picked up the initial payment for my first story, I asked where I should open an account and he suggested a bank close. No ATMs then, and my mode of living and payment was cash. I went to the bank and endorsed the check and the officer asked to see my passport, in case I wasn't who I said I was or the money wasn't mine, I guess, although I've never known a bank reluctant to accept a deposit, and when I said I didn't have a passport, he handed me my bankbook, and said, "You may draw on your balance in thirty days."

What? It was the only money I had, now in his cat box. I ran up and told Maxwell and he turned green-white, rose from his desk, and walked out. What was this? Julia stuck her head through the door and asked if I wanted a cup of tea. Soon he was back, down at his desk. "I've spoken to Mr. Greenstein," he said. Milton Greenstein, a lawyer and genial advisor on any matter, was the magazine's treasurer, yet he hand-delivered my first-reading agreement, to meet me.

At that moment the phone rang and Maxwell picked it up with both hands, so pale he looked thinner, listening, then he said yes and hung up. "That was Mr. Greenstein," he said to me. "You can draw on your money tomorrow."

Wings of Words

My life is so radiant and full it seems only natural to fly out Maxwell's door or go in a float out his window, spangled by light, to the apartment in Brooklyn Heights where my wife takes me in her arms. Words like wings weave our lives that quick. I sell another story in February, another in March, and with my third nearly ready I fly to Urbana, where she's returned to school, then she travels to the city over Easter with two friends who camp at Tony's, and I take her to *Funny Girl* and up to Maxwell's office. After she returns, Maxwell says in an unwitting matchmaker's tone, "She has such depths of sympathy in her eyes. Extraordinary in one so young."

My fourth story, what remains of "Requiem," is all but done, but we can't wait, so I fly back and she and I stand in front of a minister with her sister and a cohort from WCIA, John Ward, as witnesses, and at our motel she carries potted plants from a cloister around the courtyard to our room and sets them on the desk and night stands and then jumps up and down on the bed, and cries, "We're married! We're married! Look at all the flowers we got!"

While through all this Maxwell reminds me in a note, "I keep two one hundred dollar bills in my billfold, in case you call and say you need it"—close to two thousand in the buying power of the first year of this millennium, and one day I do call, in need, and he wires it within the hour.

He is the only person I know who always keeps his word.

Back in the city I finish "Requiem," now a twenty-page story, sell it, and pack for MacDowell. Kate Maxwell's prediction about the place proves profound! Before my time is up I fly to meet my bride in Detroit and we head north, amplifying our honeymoon, and on a newsstand in an airport find a copy of *The New Yorker* with my first story in it, and below its final columns an electrifying presence, a poem by Roethke, recently dead, my favorite poet at the time, a native of Michigan, *here*, wholly unexpected, his late light verse like a last wave from the beyond in confirmation of the story, Maxwell, her at my side—a brush of wings like the words that wove it all in one.

I fly to New York to find an apartment and she flies home to pack. And there it is, on Willow, the street I once walked, a sign about an apartment to rent taped to the bricks beside the beautiful building's door, discovered by walking Willow again.

I sign for us to assume the lease in a month.

Joseph went out every morning, after another night of awful wind, and backed the Ford tractor from the north garage and, raising the blower as high as it would go on its three-point hitch and slamming it again and again into the sandy snow to break it up, blew a path past the furnace to the house. The drifts were as high as the garage door before he backed out, and when he stood in the channel of cleared snow, its upper ridge was at his shoulders, the Ford tractor hardly higher than his waist.

Then he blew his way to the road, even though holding the steel controls of the tractor meant your hands (mine anyway) were frozen in a minute, and blew away the worst of the snow there. Our neighbor to the south, a home-health nurse who had rounds to make every day, called to thank him.

"Oh," he said, "we have to get out to get wood."

Our obsession about the winter was narrowed to that one point: keeping the furnace in wood.

I have Van's promise, so for a week we stay at his apartment on West 9th. I sit in an antique chair at a table with bow legs and hammer out on his typewriter an overblown version (I must make it *drama*, I think) of a honeymoon conversation on a beach in Michigan—shades of the other story set on a beach off the loop years ago, and too close to journalism.

A Maxwellian maxim in reference to fiction is "You can't write about something until ten years afterward"—meaning it takes that amount of time to gain aesthetic distance or you merely do cooked-up journalism. He's away for August, driving through North Dakota, to be exact, on his way with his family to Oregon, where Emmy was born and grew up. So I hand the story to Robert Henderson, another editor, and it goes through without a hitch, my fifth—a breakthrough, as I see it: the presence of Maxwell not necessary to make a sale.

Our long apartment, with a floor of oak parquet, faces east. Off the living room is our bedroom, with a window in a turret, that glows gold at sunrise. We wake to sun shining on the parquet I keep polished with paste wax and see it set on the face of the building across the street in an orange overlay. She buys bolts of cloth and starts curtains. I settle into a back bedroom, placing doors flat on file cabinets all around its walls to form surfaces to work on, and she makes me curtains of monk's cloth, my cell.

A narrow galley kitchen opens next to my office, with a dining room outside, and after each of us tries our hand at a variety of hamburgers, she begins cooking a refined French cuisine. If we want a piece of furniture, a circular table or step-switch lamp, we pay cash. We find an ornate, turn-of-the-century sideboard for a bar and stock it with a dozen varieties of hard liquor and liqueurs along a glass back I've installed.

I want to give her spiritual and artistic and emotional and physical gratification and, well, whatever she wants. There is no one I would rather talk to, no matter what (our talks sometimes last into the morning), and no one more my equal, even above me, than her. She is again to me the magnetic pole she was, which Maxwell was replacing, and he the encourager of a process she generates and has already overseen.

The couple below complains about our footsteps and maybe knocking elbows on the parquet floor, so we buy fitted carpets for my office and the entry and a Karastan runner for the living room. This room has a fireplace in one wall, trimmed with white marble, that works. As soon as we're settled, we hold a housewarming and invite everybody we know, including Van and the staff at Attention and Bob and Tony and whoever is staying with either of them, plus anybody from Illinois or Miami visiting the city, and "The place is wall-to-wall people!" as somebody yells while I try to keep up with mixing drinks and washing glasses and tending the fire—the only ones absent the Maxwells, who don't stay out late, and Bob, busy with his first shoot on Long Island.

The only incident in the uproar of noise is when the date of a friend from Illinois goes to use the bathroom after one too many and starts to fall and grabs at the shower curtains and brings them and the metal rod down with him in the tub, so it sounds like he's shot himself and hit the deck. The inner plastic curtain is ripped but the other, a fabric from Design Research, is OK, we still have it, and when I replace the metal rod, all is in order.

After the party my wife lies back in bed in her floor-length fitted gown and murmurs, "That was as good as my favorite long-time fantasy—a party with everybody I've known in my life packed into one room."

. . .

At the end of October, on my 24th birthday, after I open my gift, a silk robe, we sit on the couch and she holds my upper arm as my sister used to hold my father's. Our buzzer rings and a voice cries, "Tricky treat!" It's Tyner, clomping up in high heels, wearing a blond wig and a leather mini-skirt and blouse of Lo's, with bulbous accouterments up front, his lips painted gaudy orange, a giggling clown in his wake. He kisses us, leaving huge prints.

He has driven in from Poughkeepsie and roused Barry, our former roommate, who now lives in the East Village, and painted him up as a clown, an odd choice with Barry's big orange beard, hardly room for makeup, X eyes. I shake up martinis and we sit at the table and laugh so hard tears streak our faces, and then Tyner leans over and drops onto the table from his bra two grapefruit. "Breakfast," he says.

He stays overnight on the couch, Barry on a carpet, and before they leave I ask Paul to send "Vittorio and the Llama" for Maxwell to see. A month later the story about Herby arrives with a note from Paul saying he can't find "Vittorio." I carry the Herby up to Maxwell and in a week I hear the magazine has taken it, a Tyner story.

Bob calls to say he's bringing a housewarming gift and arrives with the most lasting and perhaps permanent of his women, Mary Ann, who wears brief skirts and net hose and does her hair in a bun and flies into such flaming arguments about films and feminist issues people wonder how they've offended her, while Bob eases back, smiling, deep tucks accenting his pleased eyes, happy with her.

He strips brown butcher's paper from a huge object and holds up his father's watercolor, *The Actor*.

"It's yours, baby!"

I tell him I can't accept it.

"What do you mean?"

It's difficult for me to receive gifts, or was then, and, yes, I know the flaws this implies, but I also sense he's been drinking. "Your dad gave it to you," I say.

"Hey, I kind of took it from stuff at Mom's! He's got hundreds. Oils! I like a lot of them better."

We have Pernod, neat, on the rocks, as I like it, a foolishly eruptive drink for persons of our constitution, the watercolor leaning against the easy chair where Bob sits. A dim silence descends, as if we're back in Jimmy's, but with no gestures left. Bob studies me, my wife, Mary Ann, trying to loosen us up with his smile, under the light of our step-on lamp above his chair, and I hear Mary say to my wife, a confidante in feminist matters, "He didn't know who Mary Shelley is!"

She means Bob. He grins, goony, tossing his head in a hoity-toity way, and I try another tack. "It's too much."

"What, I can't give what I want? You said you love it."

"I do. But really, Bob, I—"

"Hey, what's the *matter*?" He's on his feet and over me in a second. "It isn't good enough for you? You're too damn special, or what? Come on!" he cries to Mary Ann, "we're getting out!"

"Bobby . . ." she starts, but he's already gone.

As soon as she's out the door I grab the watercolor by its wire and follow. My wife calls, "Don't! You'll make it worse!" But I'm already rumbling down the stairs. All I mean to say is, All right, if you insist, I'll take it, sure, thanks, I'm sorry. They're across the street and I yell. Bob turns, setting his hands on his hips, his white shirt with sleeves rolled up incandescent in the night, waiting. His back is to a low steel-picket fence along the sidewalk, and I see Mary Ann take hold of his arm.

"Bob," I say, running up to him, out of breath. "Really, all right, OK, if you—"

"You're giving it *back?*" he yells, and grabs my shirt and spins me so hard buttons pop and the watercolor flies the length of my arm where Mary Ann, resourceful, grabs it. He bends me backward over the pickets until I'm sure its spear points of steel will puncture my spine. Mary Ann screams, "Bobby, stop that! Stop! You're *friends!*"

That does it; he lets go. We go back up and find only an eyelet for the wire pulled loose from the frame, and I accept the gift, I thank him, and it's over.

Near my wife's birthday, in December, the Maxwells come over for dinner and, besides a French meal, my wife prepares a dessert that takes hours, a blushed pear marinated in Cointreau, and when we finish and move to the living room, where Emmy Maxwell can see under our oak dining table with legs on skids, she says, "Your table is like a sleigh!"

Oh, we're on our way to heaven!

. . .

We take a train to Poughkeepsie and Paul and I search for "Vittorio" but can't find it, never do—that story like no other. In the evening the two of us sit in his darkened living room and he says, "Let's not be old farts. Have you noticed all the old farts pontificating in *The New York Times*? They were young once. Let's not do that."

"Is this a pact?"

He takes me to a gathering in an Eagle's Aerie or the like and drinks and dances most of night, looking for a girl from Vassar, he says, and on the way out, at the head of a steep flight of stairs, says, "Here goes," and suddenly is catapulting over backwards, his head seeming to remain in one spot as he spins, then down like a bowling ball, so that he hits not quite on his face, collapsing like a rag, and I see shattered white bits go flying, some clacking down the stairs, and think, My God, *his teeth!* I help to right him and he reaches to a white chunk at his feet. "My Chiclets," he says, touching his breast pocket. "Buggers flew from it."

He is working on a novel, sending pieces to Maxwell, I learn. I now know more about the ways of *The New Yorker*, or as much as one can. No way is unalterable but the route a story generally takes to get published is this: all submissions, even the unsolicited ones arriving blind, are at least looked at. Maxwell carries a picnic basket of them on the train to his country place each weekend and reads them on the way there and the way back, too, if he isn't finished up, and whenever he finds one with merit, he sends it on.

He develops such a sense about this, he says, he can usually tell from the first paragraph if a story will work, always after the first page—though he gives each one the chance of a bit more, and if anything noteworthy happens soon, he reads on, often to the end, but his first intuition is almost invariably right—"a sense like a sniffing dog's."

Once a story is "sent on," whether you're a first-timer or a contract writer, the process is the same, for fiction, anyway. The editor makes comments on a note clipped at the top and passes the story to at least one other and often two other editors to read—in my case, Henderson or Hemenway or Rachel McKenzie, usually. If they agree on the story, it goes to Mr. Shawn; if not, another might read it before it's sent "up," as it's put. That means Shawn. He reads everything that appears in the magazine, including the ads, rejecting any of these he finds in bad taste, and tends to trust his editors. But he also has unilateral power of yes or no.

Sometimes he solicits fact pieces, usually from his stable of writers, and works these through without any others in on it, but not so with "casuals," or fiction. If I hand a story in at the beginning of a week I can be sure I'll hear about it the next Monday at two, after Maxwell's nap. During the week Shawn spends his time putting the upcoming issue to bed and over the weekend reads what his editors send on.

Once when I have to hear right away, as Maxwell judges it, Shawn reads the story the night of the day I turn it in and has word to me the next morning: yes. The story is called "The Boy," the longest I've written, over thirty pages, set in upper Michigan. I'm working on the book about North Dakota, the one I once visualized, with most of my stories set there, but one day as I walk down Willow to our building I know "The Boy" and the story I wrote in Van's apartment are the foundation of my first novel—a swift one, as I see it. I tell my wife the news.

As stories appear, editors write to say they're interested in seeing a book, when one is ready, and I have lunch with the most persuasive, Joe Fox at Random House. He wants to see a novel before he considers a collection of stories, he says. Fine with me. I'm getting down scenes from the quadrants of the panoramic book I pictured during my last days at Urbana, and whether it's a novel or a collection, I'm not sure.

But now this new one has risen.

Fox says I need an agent and suggests one from Russell-Volkening who handles Heller and Puzo and Roth and has the euphonious name of Candida Donadio. I call and I'm so drawn by her dusky contralto and no-nonsense spirit I ask if we can meet. Her dark hair is in a long braid, Indianlike, her eyes darkly direct, and I hear myself telling her I'm going to finish a novel in two months. "Some can," she says, "and you just might be one of them, kid."

Then the lights go out. In our apartment and, as we learn, the whole building, in our neighborhood, and finally, as reports arrive, all of New York. During rush hour. As night falls we walk to the promenade and see only one building with lights—the telephone building, it turns out, the reason our phone works—and across the river the rest of Manhattan, its ridged and dimensional skyline, is coal black. Care gets out food, I run and buy ice, and we go up and down the stairs of our building, inviting people in, and soon a community sits in our apartment, eating and drinking by candlelight.

I'm primed to do the novel but feel I can't, given the distractions, sonic and pleasant, of the city. And it's Christmas. My wife's parents drive out to see us, bringing her paternal grandmother—first-generation immigrant from Norway—and Care finds out about a Christmas party at the Home for Norwegian Sailors, and takes her grandmother to it—an event she will mention to her death. She is from Oregon and to ease her travel (and for medical reasons) my wife's parents fly back separately, her father with his mother, my wife's mother to Chicago, and leave their car with us, a black Lincoln Town Car, lumbering boat of the kind used as a limo in the city, with the understanding that we will bring it back on our way to where I want to work.

We drive to Manhattan to see Bob. He's been unhappy since he was one of two unknowns considered for the lead in a Mike Nichols movie, *The Graduate*. But when he sees the car he responds with his altar-boyish glee. "Heyey, *ba*by!" he says "Wait! Let me run up and get something!"

He comes down somber in a black overcoat and visored police cap from Delahanty and intones, "You guys in back."

"Do you have a license?"

"Come on, come *on*. The celebs are waiting for you at the party!" We get in back, he settles at the wheel into a role, asking over his shoulder, "Where to, sir? Ma'am?"

"That party," my wife says.

At a clogged intersection he jumps out and yells, "Hey, move it, *move* it people, can't you see who I got here!"

He glides up to Jimmy's, beside a hydrant, on purpose, hops out and hurries to my side, removes his hat and holds it over his stomach, bowing, and, half bent, pulls open the door. He whispers, "Royalty, guys, remember? Walk nice!"

People at the bar come to the window, the bartender appears, and Bob plops his cap on and pulls it low. "Jeez, what am I *doing*? They know me here. You guys go in."

I'm hesitant about the car but he bows again, keeping his face out of the light, then runs around to the driver's side, and is off. Inside, with people craning to check us out, we take a booth and, before we can order, see him walk in, his cap to his ears, crown bulging, wearing my mother-in-law's cat-eye shades. "Couldn't park," he whispers.

"Sorry!" I cry in a British accent, and we leave.

He drives to a joint up Fifth, parking at the hydrant, and walks ahead of us inside, saying, "Make way. Come on, make way, they're *here*." He slides a hand in his coat as if to check on a piece. "A private booth," he says to the maître and we're in one, at the back, like that.

"Bring our bodyguard a drink!" I call, and he cries in a falsetto, loud enough for everybody to hear, "Oh, no! No, no, no! Not when I'm on the *job*!"

My wife and I leave for Chicago in January, by way of Buffalo and Niagara Falls, continuing our honeymoon. One settled unhappiness we share, a grief to her, is we can't have children. Me. My motility, as a physician in a smock has put it, is borderline, due to my slowed nature or the spate of infections or whatever, with countervailing antibiotics packed in.

In Ontario we hit the edge of a storm that dumps the most snow on Chicago in decades. I pull off at the first fuzzy show of neon, a decrepit place with an elderly owner who totes in an electric heater to bring our room up to thawable temperature—the Bates place, as I think of it. A bed crams the room, a stand beside with a pay radio on top, plus a wooden chair, a skewed representation of my room on St. Marks, only smaller.

We use the Magic Fingers, our one luxury, and while the pay radio purrs and she sleeps near my arm I jot down ideas that came during the drive. Then the novel comes tumbling out and I write as fast as I can to pin with a sentence scenes in a movie going past—a few words to open or conclude each one or give me a view into it, and when the sun rises I have a pile of notes on every piece of paper in the room. *The novel.*

My father lives in Woodstock at the present, north of Chicago, and he drives in and rescues us, once we're there, from the storm. Then gives us a vehicle to drive to a cabin Care's family once rented, where I intend to draft the novel—those scenes I visualized down in words. The cabin is where the novel will be set, in the heat of summer, but we arrive during a blizzard. By the time I get a fire going and crawl into a sleeping bag with her, I'm sure I'll die before the morning and the novel with me, if indeed I could put in words what I saw.

I try to keep silent, as older children do when they cry, and she asks what's wrong. It seems futile to try to explain the futility of driving this far, leaving our apartment, to write a book that's beyond me, if we live.

"You'll be OK," she says. "Let's do it for the heat."

We raise sweat in the sleeping bag and the next day I write the opening scene. I have to wear a glove on one hand and my breath steams like a teapot, but in two months the draft is done. We drive back in a 63 Bonneville convertible we buy from my in-laws, and in our apartment I set the draft beside a new Selectric I've bought, and there it sits, dead.

My internal bleeding has been poured onto goldenrod pages and I'm as good as gone. I start typing it up and get through a few pages that reproduce the opening of the story that set me off on the book, and the world closes down. Paging through the manuscript and seeing the varieties of handwriting strutting in an upward slant or zooming to a catharsis, some of it as staggering as my handwriting in third grade, besides a riffle of interleaved notes (one on the back of a matchbook, the backbreaker), I don't see how I can go on.

I have a book I can't move ahead so I meet with Candida of the lovely contralto and she says she'll take me on, verbal agreement only, so if either of us wants out, we're out. She asks about my name and I mention some of the variations and she leans back and closes her eyes, her hands on the table as at a seance, and finally says, "Your book has a long title with a W in it, probably begins with a W, and there's a comma in it."

"Wait a minute. I haven't told anybody what the title is—well, one or two." My wife and Maxwell, mostly.

"And don't," she says. "The good ones have a way of getting out and getting used. I have a bit of witchery in me. I can almost see your title, kid, but not quite."

2

Bob calls. "I wonder," he says, "could I ask a favor?"

"Ask."

"Could I use your car, or you help me and my dad, if you've got time, to pick up his stuff from an overseas container that's on the docks?"

We rent a U-Haul trailer and drive out, his father in the back of the Bonneville, under its slanted whomping rag top. We get a pass at a guardhouse and find the right place and when I see the big blue-metal container, I say, "How can we do this?"

"Just a little is mine," his father says, "here at the front," a hand on the container.

He is laconic, with the languid slowed nature of another culture, tall and trim, every muscle alert, from his look, with the handsome fine features of a matinee-idol heartbreaker, the wisps of gray at his temples adding an aristocratic elegance.

We get everything in except several medium-sized oils, vivid, erotic, too precious to load first and now too large to pack in the trailer. Mr. Senior studies me with his muted distance of irony, the take of a movie actor if I ever saw one, and says, "You're right. We can't do it."

"The top," Bob says, and I understand. I lower the top and they sit in the back, holding the paintings between the seats, and I creep all the way to his father's storefront studio in the Village, bearing an erotic brilliance jutting up over my head into the New World.

Paul calls to tell me his novel is done. It opens with the story from *The New Yorker*, he says—which has hardly changed from the version I heard in Scouffas's workshop—and its title is *Shoot It*. It's been taken by Little, Brown.

When I tell Maxwell I can't work he doesn't say as he did when I told him I had writer's block, "Go home and write five stories this week." I had to repeat to him what I had said, in case he hadn't heard right, and he responded with a somber stare: *Do it*. So I went home and tried, as if pushing my pencil through clinging presences; wrote the first, went on to the others to finish, and the last came easier. Then I had five stories to work on, no more block, and sold three. Now he says, "You need to think about what you have while you take time off. Rest."

He is going to France with his family and offers us their country house for six weeks. The only condition is I must keep his rose bushes, dozens and dozens planted across their rambling yard, pruned and trimmed. It's the time of summer when the bushes are going off bloom and I have to handle that in a certain way, as he demonstrates during a visit we make for practical instruction. He wears faded trousers and a cotton shirt with the collar open.

As we're about to leave, he shoves his hands in his pockets, as if reconsidering his decision to let me take over, then ushers me into his office, a small room in a front corner of the house, with an upright piano facing its door. The desk to the right is a piece of plywood large enough for a typewriter and manuscript. With the chair at the desk the two of us have just enough room to turn to the back wall, where crammed bookshelves step to the ceiling. My eyes go to a filing cabinet beside the desk and he says, "All writers are snoops. I am. I remember coming across one of my host's love letters, at the back of a roll-top desk and, oh, how I read! I'll only say what I said to Emmy when we were married, which is, 'You may read whatever you want but you're responsible for your reaction to it.' There's nothing I can do about that."

So my wife and I are tenants of their heavenly house, set in the greenery of untilled countryside, with rough woods at one edge of the lawn. The other side to heaven is the heat, "hotter than the hinges of hell"—a saying of Maxwell's father that he likes to repeat. In the inferno greenery steams. Each day after my pruning, I run around the house, over and over—trying to burn off pounds, since I've ballooned to 140—but also burning up energy simply to sleep: that horror when no words arrive.

We learn that different areas of the house are cooler over different hours of the day, and migrate from room to room, ending near the entry at night, for the record player and its company: *Otello*. In every room except his office and the master bedroom I set up a place to write. I sit in Brooke's back room and stare at the manuscript or a blank page till I'm reduced to sipping vermouth, then wake in the morning feeling my eye sockets are scorched metal tubes six inches deep.

I pick up a newspaper or a book and examine each sentence word by word and think, So that's how they do it!—by which I mean arrange words in a coherent sentence.

We sleep in the room of Maxwells' live-in maid of many years, Ellie Mae, a semi–guest-room with twin beds, not the master bedroom. The place where Maxwells sleep seems unapproachable. The kitchen is outfitted with a French gas range of the kind my wife has seen only in illustrations, as if installed for her, and books by Adele Davis. Davis on diet seems to my wife so radical yet commonsensical that she writes and asks about the prognosis for us having a child, with my, um, motility, and Davis actually answers and suggests a daily dose of Vitamin E that to most might seem fit to kill a horse, but now is common.

So my wife dollops out the dosage and I oblige, passive in my stupor, knowing that without her in one or another of our musical twin beds, drawing me from the inwardness I curl toward as to a wound, I wouldn't survive.

In Maxwell's office, which I visit mostly for books (though now and then I sit at his desk and imagine how it would feel to be him, *to write*), I find a first collection of stories inscribed with fondness from Harold Brodkey, and in a literary journal a Maxwell story about a young man on a tramp steamer in the Caribbean, never collected in a book, though I like it as well as any story of his, and a box that once held a ream of cotton-content paper but is now filled with letters from "Jerry" Salinger.

When I see what I've found, high on a shelf at the back of his office, I look over my shoulder like a kid with his paw in the cookie jar. Then I remember Bill's comment about writers as snoops. The first letter contains a hand-drawn map by Salinger to his place in New Hampshire, along with written instructions on how to get there, as he anticipates a visit from the Maxwells, and I'm not able to read too much beyond that. I may be a snoop but my snoopery has limits (I could not do as students of mine have, page through letters and bills on my desk, reading) and to look any deeper into the Salinger seems, more than snooping, intrusion. I stop.

I do notice, however, that Salinger is giving religious advice, especially to Emmy, about adopting his way of Zen. About *The Chateau*, in which a couple coincident to the Maxwells takes a trip to France, Salinger writes, in effect, "Hurry and finish it, please. I can smell their suitcases!" Oh, shades of "Franny," I think, but can't go on.

The week they plan to return from France we scurry back to our apartment, leaving notes of thanks in every place in their house they might look—except the box of letters, I think. Once back I feel as bleached and ragged as an old sheet that's accommodated one too many—the stream of characters that has poured from me in three years, a dozen stories sold, a dozen in process, a hundred or more pages on two different "first" novels, and a first novel finished—its goldenrod pages already fading from the sun.

I set it beside my typewriter and this time, seeing the minute or exuberant or sometimes staggering handwriting and the interleaved notes, I'm amused, and then, staring at the first page, a new sentence arrives: *It was close to the end of the day*. I set it ahead of the first sentence that stands and the book lifts off, I can see through all its scenes to a new light (with darkness descending) and I'm back on the roller coaster of being a writer again.

Or maybe merry-go-round is more apt, because each day revolves the same, day after day, rising late and sitting at the typewriter, sometimes in my shorts or silk bathrobe, and working all day and through the night, with breaks for meals, or I see a sandwich at my elbow and know she has set it down and get up to thank her and we end in bed, and after she falls asleep I get on my carousel horse and when I see light at the front windows, through the swag of Austrian shades, I give out.

I dress and go to the deli at the St. George and get a half-quart of Rheingold and come back and stand at the front window and sip as I think of the day's work and what lies ahead, gingerly entering the outer realm of dreams while still on my feet, and in that state fall into bed. I wake late, sometimes in the afternoon, to the sound of her busy in the apartment or to silence: *she's at the health club*.

So it goes through the second draft, then the third, on white paper, though there must be a pause between because we're running low on funds and I get two early chapters into story form and send them to Maxwell. I sell one, enough for the time. Then the greater pause of surprise, nearly enough to stop me, when we learn she's pregnant—and so the quiet concurrence between us, our hush of awe at the productiveness that being productive brings. We will be parents, with a bow to Adele Davis and the diet Care demands, and now she's writing a book on the world of a pregnant woman, sitting cross-legged, in one of her Yoga postures, at a low bookshelf at the front of the living room, under the pair of windows facing east with their swagged Austrian shades.

"I need to ask another favor," Bob says on the phone.

"Go ahead."

"My dad is having a show and needs to get a lot done quick, and I wonder if Care could be a model for him. He can't pay but he'll let her have any charcoal she wants."

"Gee, I don't know. She's pregnant, Bob."

"You can't tell and she wouldn't have to sit that long. He likes some moving around, action stuff. Mary Ann already said she'd do it but he needs another woman."

"I don't know. We've been at Maxwell's, and—"

"*Heyey*," he says, and I can hear his smile. "They keep their clothes on! That would be kind of funny, huh, to let Mary Ann do that—you know, no clothes—with my *dad*."

My wife calls his father to set a time and asks what she should wear and he says, "I want to see your arms."

She goes to his studio, and then each day to the pool at our health club, to swim x-many laps, and once a week takes the subway to an organic grocery in mid-town Manhattan she's found and orders a week's food and returns; and later, when our buzzer goes off and she answers, I hear from the tinny speaker, in a comic miniature version of the Slavic laborer who delivers for the store, "Ore-*ghan*-eek!"

That is our password for the time, from the fall of 1967 through its winter and into the next year: *organic*. We're setting out on the life that thirty years later we hope to incorporate to the full, with not only a garden and crops and animals but the land itself, for a half mile in every direction, organic.

In the cold of this office, when the wind is from the east and travels under a door not far from my feet, I feel the frigidity of the kitchen where I wrote *What I'm Going to Do, I Think*, and as I turn and lean to the desk to get down a thought that has occurred, I realize that half the thought isn't in words but an inner collusion of imagery and faith in intuition, only partly related to language.

So I sit and try to translate my thoughts.

February 29 the novel goes to Candida. She sends it to Fox, promising I'll hear in days, but by the end of the week with still no word we rent a car and drive to Montauk, to a resort we discovered on another trip, and take a first-floor room, with a deck you step from to the beach—that herring smell of sea. I read a feature in the *Times* about a group called the Polar Bears who swim through the winter in rubber suits in the variety of waters of the boroughs of New York, and after two days with no word on the book, though I call Candida and Candida calls, I decide I might as well be a Polar Bear. But no rubber or other suit. I run in black briefs in a spattering entry into watery sand and when the next wave curls white at its top, letting its tonnage down in a roar of collapse, I dive.

And feel I'll pass out from the shock, like a mallet to my gut, my balls squeezed BBs speeding toward my brain. I gasp, still under and rolling, sand peppering me like hailstones, gagging, then my feet hit and I shove myself, an erupting dolphin, above it and roar. Then worry I won't make it out but do, hunched like a beaver but bare, weaving through sand that feels hot, up to our room.

She stares as if suffering my shock: "Was that wise?"

"Egglirating." My mouth is so frozen I can't say the word. At dinner, held in a separate building, we learn the water is thirty-eight degrees. The next night she stays in the room. She's not ill but worn to weariness from carrying a child, carrying her book, carrying mine to completion, carrying me, and still no word. Monday, Candida says. I eat in solitary inwardness, though Care is near enough, and decide to order a meal and take it to her, not just a cheapskate's way of avoiding room service, but for *her*.

When it arrives, with stainless-steel covers and the rest, I see I'll need three hands. The owner motions to a hostess, an Oriental who isn't Japanese or Chinese, to my eyes, but perhaps part Polynesian, and as we go down the dark path to our building, already I'm worrying about her tip. I usually give fifteen or twenty percent, never sure which I will, but ten for bad service, a rebuke, and suspect that other duties fall into different categories. I pause in the moonlight on the path and her upturned face glows.

"What do you expect for—" I resist *tip*, then *gratuity*, suspecting another word won't arrive in my word-addled, novel-

apidated state, and she takes my arm. "For you, nothing. But with a wife you take another room. Unless . . ."

I'm so appalled I tip her twenty percent.

No news Monday, no money, so we head back. The next day Candida promises word by evening as I stare at my wife face-down on the bed. Late Wednesday she calls and says, "Are you sitting? I'm giving it to you straight, kid, and these are his words, 'I don't think this is a novel.' "

"*What?*"

"He said, 'I'm not sure what it is, but I don't think this is a novel.'"

"What the hell does he think it is? How could he say that?"

"They all make mistakes, don't be surprised, I see it all the time. It's a good book, I said that, and it's going to get you a lot of attention. Now where?"

I call Maxwell and he says, "Send it to Bob Giroux at Farrar, Straus, & Giroux. I've felt for a long time they should publish your work but didn't want to interfere."

I agree. I've begun to notice FS&G's writers and, just as much, the artistry, the physical craftsmanship, of their books—dust jackets, design, paper, covers, binding, all. But another wait; Giroux is out of town, his reader in the hospital, and by now I've cleared and emptied the bar.

Tyner calls and says he sent galleys of *Shoot It* to Mailer—Paul reads *Esquire* and admires Mailer and Dwight Macdonald—but hasn't heard back, and could I follow up somehow?

I'll try, I say, and send Mailer a letter and days later, on the way to the St. George, I see him walking toward Willow, waving away a young man with three cameras dangling and another going off in Mailer's face. "Get!" he yells, the first I've seen him upset, and then shakes his fist and utters a French phrase I know isn't flattering.

He spots a gypsy cab close to me and sprints for it, the young man in his wake, cameras flopping, clacking away with one, and as Mailer comes up, angry, I trot along with him and ask, "Have you had a chance to read that book?"

"I have to shake this guy," he says, climbing in the cab and pushing the paparazzi away with a foot, "I'll write!"

His letter begins, "This may freeze the very crystals of your piss, but . . ." His secretary, his poetaster also, he calls her, has looked over the book and feels it's fine but right now he doesn't have time to read it himself.

The night of Updike's reading at the Y, Maxwell waits with Emmy at the exit, hat in hand, and Updike hurries up with his topcoat climbing his head like crumpling wings, to slide it down his arms. Maxwell apologizes for not being able to attend the party. Then he turns to me and says, "I nominate Larry and his lovely wife to attend for us."

Bob, standing at my side, nudges me.

In an apartment off Sutton Place we enter a clamor like a blacksmith shop during futurity and find Updike, in his navy-blue suit, backed against a light-blue wall beside a pedestal table sporting a pewter vase of wheat grass, a Vermeer, with so many cameras going off he blinks even when his head goes back like a horse with a sneeze to laugh. The crescent of bodies hemming him thickens and mills, no chance to approach. I finish my drink, served in a stunted glass the circumference of a telephoto lens, and squeeze into the crowd and hold it over my eye like a lens focused on him, twitching a finger when flashes go off. Updike looks into its base, seeing what's up, and grins.

He appears to me to want another drink and I'm about to ask what when a woman holds one high and he reaches over heads for it. Not long after, he appears to have to head for the bathroom, that leveler of paper gods, and I maneuver to his side and back into bodies pressing behind, drawing Bob close. "John, this is Bob. De Niro. Bob, John."

"Hey, I wonder if there's a movie option on *Rabbit?*"

"There is, but I'm not so sanguine about it."

"Who's got it?" A whoop rises from the clamor, and Updike makes a sidewise check, eyes rolling white. "A fellow named David Wolper."

"Sure, Wolper, for how long?"

"I believe it has years to run." Updike's wife waves.

"Well, when it comes free, could you—"

But Updike is off, making a misstep of the kind we all regret, for this must be one of the missed artistic moments of the decade.

Bob calls one night and I invite him over; it's late and Mary Ann is working, waitressing at an all-night club. So Bob and Care and I sit in the front room and work on a bottle of Dom Perignon. Toward morning Care says we should have breakfast in Paris, and the next I know I'm on the phone calling airlines. I make reservations to get us there for our *next* breakfast, a day away, and Bob asks about our passports. *Passports?*

I have to call back and cancel.

Bob Giroux calls. "It's a wonderful book. We want it for our list next spring. Can you come up and visit?" At the Union Square offices I meet Giroux and Michael di Capua and an editor named Harold Robbins ("Yes, I know," he says, "and, no, I'm not him.") and the debonair owner, Roger Straus, of an old New York family, his combed-back wavy hair like a steely ocean set to break. I follow as he strides down halls in a tailored summer suit and, oh, these halls are like *The New Yorker*'s! with rooms and cubicles packed like a milliner going out of business. He uses language like a fishmonger, a pirate, and I feel the radiance of my first years with Maxwell rise in my face: *home*.

Giroux takes me to lunch with Michael and suggests I break one chapter into three sections instead of two, to keep it consistent with the rest; and the pedantic absurdity I meant to suggest by foreshortening it, a form of foreshadowing, seems gratuitous. They ask if I have further changes and when I say, I think, they smile.

"There's also an epithetic string that maybe can go."

"'Thank you very holy—'" But in the face of Bob Giroux's great dignity I can't go on. "Etcetera," I add.

His sudden smile carves a dark curve and expression gathers at his nose and around it, then spreads to his eyes in an incendiary flash. "Yes!" he says.

But ultimately my actual editor and I agree it stays.

. . .

I carry up my changes, wearing the worn white pants from the suit I wore when I bowed to her, and a striped Beatles jacket, with a belt at its waist, which I found, of all places, at Abraham & Straus, founded by Roger's family. The getup once fit right but now I bulge from it, up over 145, though I've been working out with weights—or because of that. Michael takes me to lunch and in the street notices my waddling unease. I see his quick glances, mostly rearward. At the meal he says, "I've been looking over your changes."

"You're my editor?" I study him as if for the first time: his compact face with heavy eyeglass frames exerting a tug at his features, sliding down oily rosy grooves nipping his nose, nearly bald on top but with thick dark hair in a tonsure from there down, his forehead shiningly protuberant above his eyebrows, rounded at each upper corner, as if the complexity of his thought requires extra room to revolve.

"Who did you think?" A tender prim smile.

"I didn't know. Maybe Bob, the Harold I met."

"Please."

"Bob gave me the first advice, about the chapter."

"Maybe you'd rather not work with me. It's a mighty rifle-packing, he-man book you have there. Maybe I'm gay."

"So?"

"Nothing, if you can be that mild about it."

On a walk on the East Side I notice St. Vincent Ferrer, and say to Care, who looks tired, "Let's go inside." The two of us are alone in its cool vaulting, our footsteps echoing back, and we walk into a tasteful side chapel, octagonal, mostly bare but for a marble font, and turn to one another: *Baptism? Baptize our child here?*

"Should we?" she asks.

"What do we have to lose? And who knows the gain?"

3

The advance for the novel is less than half the amount I receive for long stories, and though the book has consumed my time for over a year, I want to do another. I meet with Bob Giroux to talk about a contract for a second, the one I have a dozen pieces for, and as we sit in a seedy Longchamps on lower Fifth, he keeps looking past me across the dim room. "I can't get over how that woman at the far table resembles Eudora Welty," he says.

After further talk he says, looking over my shoulder, "Extraordinary how she resembles Eudora!" I'm tempted to turn but don't, as my mother taught. Then he says, "She's paid and now she's leaving and I'm *sure* it's Eudora!"

I get up and follow after a woman with cropped gray hair and a widow's hump, as it's called, already out the doors. "Miss Welty," I say, and she turns under the blue-striped awning in a dapper schoolteacher's suit of gray gabardine, blinking against sunlight that hits even here, her head slung low. I introduce myself and say how much I liked "The Demonstrators," which appeared earlier in *The New Yorker*, as if in answer to the King assassination, and tell her it seems to me that the poor doctor, the central character, has the weight of the world on his back.

"Why, thank you!" she says. "How kind of you! And I liked yours about the boys at college and the grandpa's dying." The new "Requiem." "Particularly the end." This surprises me because I thought the end was daringly erotic.

By now Bob Giroux has settled our bill and I stand to the side as the two talk. Then she has to hurry to make an appointment, she says, and we turn the other way, toward the Farrar, Straus offices, and Giroux says, "Thank you for intercepting her. I haven't seen Eudora in years. She was with me at Harcourt, you know." But I'm inside another summer, with Shattuck and his wife lying on their lawn, his wife with *Golden Apples* open on the grass, and how my love for him never dimmed.

At the office Giroux asks if I want to attend a performance of Lowell's *Endecott and the Red Cross,* maybe have a drink with Lowell afterward, if he's free, and I say, "Only if my wife can come."

Giroux and Michael and Care and I wait at a table in a restaurant-bar for Lowell, and Michael says to Giroux, "It seems to me that Robert is maybe in a manic phase."

"Oh, no," Bob says. "I don't think so. I think he's just extraordinarily happy right now. It's wonderful to see him so, don't you think?"

At Illinois I recited poems from *Life Studies* for a literary conference—"Skunk Hour" and "Waking in the Blue," both homoerotic with religious overtones. Sex for me then seemed a religion, as in "Skunk Hour"—"I hear/ my ill-spirit sob in each blood cell,/ as if my hand were at its throat. . . / I myself am hell;/ nobody's here—"

"There are no Mayflower screwballs in the Catholic Church," he wrote in "Waking" (for a time a convert to Catholicism, like Hemingway; both saw spiritual hope in it), and I have heard stories of how Lowell takes on the persona of Jonathan Edwards and preaches such Calvinistic sermons that the name Cal seems to refer more properly to John of Geneva than to his original namesake, Caligula—that role Jay Robinson played in the Hollywood spectacular *The Robe.*

My wife and I attended Lowell's reading at the Y after *Near the Ocean* came out, and I was dismayed by his voice, fumbling yet genteel, all thumbs, with vowels taking the shape of his years in the South with the Agrarians, as if the accents of Sewanee rather than Boston had seized his tongue and a muted Easterner was trying to find his way out.

Lowell hurries into the restaurant with a scarf dangling and I stare at his right hand, closed in a fist on the table's edge as he says he abhors the conventions of theater. But he seems buoyant and alert, glancing with sharp eyes at Giroux and my wife. Bob introduces us and Lowell gently shakes my hand, then nods at her. She soon has to leave, worn out by the hour, after a play with no intermission, so I walk her to the corner and hail a cab and get her inside but return feeling uneasy.

She looks tired beyond mere tiredness.

Another poet has mentioned Lowell's "shark smile" and as Lowell continues in his exuberant vexation about the play, the smile overtakes his face, drawing up a slitted bow, exposing rugged and regular teeth polished by words. And I can tell scarlet billows will flow with a nip ("He *is* on a high," Michael whispers), as I watch him lean and nibble at a vodka or a gin; I didn't see him order but the liquid in his glass has a viscous cling.

I've just read Mailer's *The Armies of the Night*, done through Aquarius, as Mailer calls himself, and as Mailer wound into his sentences with their classical tang and went into detail about the march on the Pentagon, I was startled to find in their curious wrap Lowell, the person, arrested with Mailer and others by the D.C. police, and as soon as I have a chance I ask what Lowell thinks of Mailer's handling of him, suspecting Mailer's hopped-up prose and guesses about Lowell's thoughts are an affront to his dignity.

He hikes his glasses higher, their lenses so smudged they seemed to steam in nearness to his unblinking stare at me, and says, "I think Norman caught something profound about me"—turning it on me in a playfulness that has the edge of a knife, that smile. "Don't *you?*"

This is to Giroux. "Yes," Bob says, and that's that. Then he starts on Randall Jarrell, another poet published by Farrar, Straus & Giroux, and with Giroux next to me, with his yellow-white hair that looks powdered and his flesh packed so stoutly in his skin it seems the source of his rosy complexion, I understand how a name in print can abstract a person, so that what registers is a book or a business or "reputation"—this presence now more florid from drink, saying how much he dislikes the talk about Randall going around, three years dead.

They all seem to agree, Lowell, too, that Jarrell was the intellect of the Lowell-Roethke-Berryman generation; then that daring plainness of everyday speech in his last poems. Pure poetry or sugared talk were the opinions. Before, his poetry was packed with a recent pursuit, the Air Force or opera or painting or the German language or sports cars—America's poet-critic, on a level with Eliot, but wittier, with a caustic streak: a line about one anthologist was that his verse sounded like it was written on a typewriter, *by* a typewriter.

Jarrell seemed on top, dapper and winking, light years beyond mere writers, but after his book of plain speech he had a breakdown, and not long after, once out of the hospital, was hit by a car as he walked along a highway at night. Michael was responsible for moving him from Holt and sees him as *his*, and now he says it was a terrible accident. Giroux concurs, though he's a rapt listener likely to side with the one speaking, and we turn to Lowell, Jarrell's former college roommate.

"Randall wasn't looking for the car, I would say, and I won't say he walked into it, but could he have sought it? The driver said nobody was in sight and then Randall was in his headlights, lunging. I'll always wonder if he jumped."

"Please," Michael says. "He'd been walking at night for, well, a long time. Trying to get out of the way can look like a lunge. He had too many good books going to do anything so stupid!" He looks angry and I'm uneasy at his peevishness with Lowell.

He says, "The thing about Randall, more than any poet, is when one saw a new poem of his one had the hope of encountering a masterpiece. He was the one to do it. I don't believe he thought his ego had the audacity to act in a way he did not want it to and go cracking up on him. What was worse, once he was home, was that Mary, his wife"—Lowell seems to add the tag for me with a quick glance—"kept calling him a child, because he wasn't getting over the breakdown."

"Where do you get *that?*" Michael asks, incensed. "When I talked to Mary she was concerned, she was *kind*."

"From Randall is where I got it. I called and he said, 'Cal, Mary thinks I'm being a child.' He knew I'd been through this so many times he could talk to me. Imagine calling Randall a child! That disdainful, aristocratic dignity of his! You can't describe him so without its having a deleterious effect."

He suddenly turns to Bob, with a comment about Peter. "Peter *who?*" I whisper to Michael. "Taylor," he whispers, staring at Lowell, livid.

"I *wish*," Lowell says, "I wish Peter would stop wasting his efforts on the plays he's been writing and turn to a novel. That luminous fiction! How I miss it!"

Then he's talking about John—"Berryman," Michael gets in sotto voce, and I know that Bob Giroux, Lowell's editor, is Berryman's, too. Lowell says that a few days ago he visited Berryman in a hospital where Berryman is committed—to dry out, apparently. "And there is John, lying back in his bed, the normally clean-shaven John, calling to me from bed with that beard like Falstaff's—*like* Falstaff, too!—'Come, Mistress, Quickly!'" Shark laugh and a swoop of his smudged glasses all around. "He had only a mat on the floor in the middle of the room. He pointed a shaking finger where I was standing and said sadly, dead flat, 'There I meditate.'

"I moved myself from the spot. 'Saul was in to see me yesterday,' he said, and I suspected Biblical visitations. But Bellow. 'Saul encourages me to meditate,' John said. 'Oh,' I said. Little conversation after that. When I was leaving he said in the same flat, dead, sad-eyed way, 'Cal. Damn you. Bring me some booze.'"

This seems a slap to Giroux, who recommended his dry-out and is on his feet, slipping one sleeve then the other of his topcoat on, arranging a scarf as he holds his white head at an odd bowed angle. Lowell gets up and the two are off, Giroux looking like a raconteur of the literati, needing to hear more, his head inclined to Lowell.

Out on the street, Michael and I turn to each other in the sour dead-end of the night. He grips the handle of a plastic bag of books, copies of Lowell's trilogy of plays, *The Old Glory*, I figure, still angry. "No wonder no wife of his can get along with him. Talking about Peter like that with his play running! About these poets, *really*, sometimes I think, the way their wives must, Who wants to put up with a mean drunk? Vocational simplification. Ugh! You see why I don't write. Bye."

Before the opening of De Niro Senior's show he calls to ask if my wife will attend in the dress she wore when she posed. The Zabriske Gallery, where his show is hung, is on Madison, and when we arrive we mix with Bob and Mary Ann in the crowd. Then I walk around the several walls and look for a watercolor of the kind Bob gave but see only charcoals and paintings. Then there she is, over and over in her elegant poise, with its hint of distance and languor—caught in every line as in the charcoal she chose. The rapid yet unassailable lines as in *The Actor* are even better, a subtlety mastered—Care saying if they didn't fall right in the first strokes, he not only tore the sheet off, he balled it up, then tossed it across the room. His colors have changed, the muted irony with bright swaths now, the emphasis falling in compositional chords, and most of the paintings, especially the portraits (one of Mary Ann and my wife posing together) convey a sense of ascending.

Afterward we walk from the subway to a restaurant in the Village for dinner on Robert Senior. *His* father is with us, tall and heavyset, silver-haired, with a way of walking with his arms out from his sides, as if about to quick-draw on the louts he sees on the street, as he calls them.

"Bobby! That's him!" Mary Ann yells, pointing to a dapper dude in a jacket ahead who turns with the look of swallowing a mouse. "That's the manager, the guy that's always feeling me up and saying he's taking me down!"

The man tries to run but Bob gets him by the jacket and throws him against a building, then socks him in the side of his head. "Kill him!" Mary Ann screams, but Bob's father and grandfather have hold of him. "Don't let him get off like that!" she cries. Bob struggles with two generations of De Niros and yells, "You touch my girl one more time and I'll bust your ass, you fucking scumbag!"

The man, holding his face, says, "I'm calling the cops."

"Do!" Mary Ann says, "and I'll tell 'em where you're always sticking your hand—scumbag!"

"Hey!" Bob's grandfather says, then he and Robert Senior get firmer purchase on Bob, and the brawny elder man says, "Bobby, you cannot do that in this city. Let's go!"

The dinner is over, the night is over, and Care leans against me, weary, then lays her head on my chest.

. . .

We can't decide on a name. We run through names of relatives, page through naming books, the city phone book—nothing. Then in one I find Newlyn, Welsh for "new spring," and Care and I stare at one another as if in recognition. We decide to add her mother's maiden name, Smith, to counter the poetic tinge, and none too soon. Two days later my wife is in labor, six weeks early, our first child premature.

May 15, 1968. I believe I've recovered enough now, physically and emotionally, to speak of the birth of our first child, our daughter, Newlyn Smith. What a beauty! And now I know, as I've told several people, including Maxwell, that a girl is what I actually wanted. I knew it when I saw her and I'm completely at ease when I touch or handle her. A boy, I think, would have been harder; I always would have been worried about the relationship, and if I was treating him right, and hung up about his future. No such problem with Newlyn. She makes me feel at home.

1:30 A.M.. Newlyn just woke and I'm sitting here holding her while Care, who's very tired, tries to get some rest. Newlyn is now on about a two-hour feeding schedule so it's hard to get a lot of sleep. Newlyn quiets in my arms, usually. I have her here in my office. She's nearly the size of my typewriter. I hold her head in my cupped hand and her legs don't reach to my elbow, quite. That small, so early, five pounds. Orangy-brown color to her skin in places, and the texture of an orange from subcutaneous oil, such a beautiful, expressive face. Long, perfectly shaped thin fingers, each hand only slightly larger than my thumb. Her hair a light brunette (there's lots of it) and extremely fine, like her mother's. She opens her eyes and looks at me—eyes just beginning to focus. Now she's sleeping, one hand lying on my chest, the other over her right eye. Her hands and arms are already graceful in their movements, and sometimes her gestures and expressions are those of a grown woman. It's still nearly incomprehensible that she's a life, a separate life, a new life, a life from her and me, but already her own person. As I write this, she stirs, placing both hands, fingers held in, on either side of her head. Her mouth, shaped like a crossbow, is slightly ajar. She just made a sound, perhaps questioning that crossbow image. Such shapely, delicate lips. Her nose is a Smith nose, her hair is half Care's, half mine. When she wants attention and cries out, her voice is like a girl's—high, pure, feminine, imperative. She'll be a singer or dancer—or whatever she wants. She's charmed everyone who's seen her . . .

To be continued tomorrow. She's sleeping well and I'm tried. It's nearly 3:00 A.M. *I've bedded her down in her vanity drawer bassinet.*

She has arrived so far ahead of schedule we have no crib, though the Maxwells are giving us theirs, and she is anyway too small for a crib so we pull a drawer from my wife's oak vanity with its tall oval mirror and pad it with soft bedding, and she sleeps within reach of our bed.

She was born May ninth and on her second-week birthday I place her at the end of the couch and prop her head with a pillow so she's half sitting, half lying, and see that she's put on so much weight, especially in her cheeks, she resembles Edmund Wilson.

We gather in the living room, all five of us, or the five who remain, Laurel and Ruth and Joseph on the couch, Joseph's leg out and a crutch at his side from the injury he's nearly through, my wife in the comfortable reading chair, I in our deacon's bench against the wall, and Joseph says, "I'm never going to do that again."

It's April and we are through the worst of the winter, with the melt and runoff so heavy it came in a cascade on the greenhouse floor, then a smothered gurgle as it rose two feet up its walls on the same day that fifteen houses were flooded by the Cannonball River in Mott; and now waters are gathering and spreading into a sixty-mile–wide lake that is spilling across the border into Minnesota, burying areas of Grand Forks, part of the monumental flood that will soon spread internationally, as it were, over the Internet, with new photos of it available every day.

So we've decided to go on a family outing, to see *The Hostage* at the Playhouse in Mott, and when we walk out we find snow piled higher than a top hat on the hood and roof of the gouged Lincoln, a snowfall arriving so thick I have trouble seeing the lines at the edge of the blacktop and then on our gravel road nothing, a white-out for miles, Joseph watching for the ditch out the rolled-down window on his side, I on mine, sometimes sticking my head out, and if it weren't for my hundred trips down this road, no, thousands, over four thousand I recently calculated, on my way to banks and grocery stores and parts shops over twenty years, so that I know the road by feel, we never would have made it back, never returned, and so Joseph says, "I'm never going to do that again, no matter what."

My head hangs so low I feel I can dribble it over my knees, and finally I say, "Let's give thanks."

. . .

I sit in Bill's office and relate how the obstetrician arrived at the hospital after us, Care already in the delivery room, no time to prep, as they say, and how he threw me a gown and paper cap and shoes—enacting this in my new incarnation as Maxwell's Jester, entertaining him with accounts of my outrageous statements or skewed situations—and how I'm shunted into the student auditorium rather than the delivery room, because it's happening so fast, with raked seats running to the ceiling and Plexiglas to my eyebrows, but I keep yelling the instructions from Lamaze to my wife anyway, my paper cap falling off, while our OB sticks his head in and out a side door to direct the nurse who happens to be handy, meanwhile trying to scrub between giving directions, and runs in just in time to place Newlyn on a scale above the table, as if that's most important, then turns to give more instructions, so he doesn't see that Newlyn, crying, protesting, has a hold of an edge of the scale with a miniature hand as if to pull herself out, the swinging scale starting to tip as her head rises, and I have to yell "Hey! Hey!" from behind my Plexiglas barrier, "The *baby!*" He leaps back in time, grabs her, gets it steadied, gets her weighed, and I yell (or so I say), "Next the mercury stuff in her eyes, huh?"

"Oh, goodness!" Maxwell says, tears spangling his lashes, and wipes a curled little finger of each hand under each eye quick. "You are something!"

I call Lloyd, quarterhorse breeder Ruth has worked with as a trainer, and tell him the tale of the tails.

"That's not from the cold. All my horses, except the few at home, are out in it every day, without a barn to hole up in, and their tails are intact, I can assure you."

He was once a science teacher but traded that for mornings on horseback under the blue.

"Then what could it be?"

"You don't happen to have sheep with your horses?"

"Yes we do."

"They ate them. It's inevitable when they're together and their feed is low, as I was warned years back—something about the protein in horsehair."

Before I can make excuses, he says, "This is the worst winter for feeding I've seen, and I'm in my sixties. And my father, who's considerably older, of course, says the same. Your horses are quiet and easy, the only kind to have, and probably hardly noticed, maybe kind of numbed by the cold. They'll grow back. Greet your Ruth for me!"

In March I have to speak at a home-school conference in Bismarck, and my wife and I take Laurel—after a winter with fifty people in the house, as it feels, on one another's backs. In the morning before my first talk the phone rings and my wife gets it. "Yes," she says. Then, "Oh, no." She holds the receiver to her ear so long in silence I know it's serious. Then she drops it into the phone base, staring off. "That was a nurse. Joseph shot himself."

"*Shot* himself?" I say, and stop. Because in this fractured world of slim hope for the young I suffer the worst of a parent's fears, *suicide*, though I know Joseph—

"He's all right," she says, and turns to me as I hurry to take her in my arms, then cries, "It's his *leg*."

In the story as it unwinds, with details later added, what happened is this: Joseph is protective of his sisters, maybe overly so, and when he is home alone with them (this time Ruth) while we are away, he will load, as he did, a clunky six-shooter he bought on a lark years ago, a .38 special, and keep it under his mattress; when he wakes in the morning and picks it up to unload it, an announcement on the radio downstairs catches his attention and he steps into the hall between his room and the girls' and strains to hear as he raises the pistol to swing its cylinder out and spill the cartridges, steps again as it slips and drops behind him. The safety has never worked, the hammer is faulty, and Ruth is jarred awake by an explosion outside their door. She hears his shout, then he says, "I'm hit."

He's on the floor and as she runs for towels he gets up and grabs the pistol and disassembles it and tosses it under his dresser, so nobody can use it, and by now Ruth is trying to get a towel around the bleeding swimming from his knee. He gets down stairs on his own, out to the car, which Ruth has trouble getting to the house—the snow—and she drives him to the hospital in Elgin. There the doctor orders an ambulance for Bismarck and he is on his way now, will be a block from us in minutes.

"Dad," he cries, rolling in on a gurney with two men trotting beside. "I'm sorry we don't have wood!"

I have to restrain tears that burn like alum to keep from shouting, *My God, forget about the wood!* But it's been more than an obsession, it's meant survival. I dig to the hand I see at his side below covers and take it in mine.

"Joseph, I feel so bad. You're so dear to me. If I could take this on— You can say at the Air Force Academy you know you can take a hit. Most can't. I want—"

The attendants attempt a turn, rolling the gurney toward a pair of doctors calling from a room, impatient with me, and I try to follow and then have to release his hand, that parting. He goes, gone from me, a son on the route all children take to adulthood, their father or mentor in the violet of shadow. Though I want to say, *You're in me.*

"I'll pray!" I cry.

His mother won't let go. She stays on at the hospital after we hear the results of the x-rays: the bullet went through his upper femur and clipped an edge of his kneecap without hitting a vein or artery or nerve or tendon or any primary muscle. "Nobody could do this if they planned it," the black-bearded doctor tells us. "It's a miracle."

He has asked to see me. So I go to his apartment in the East Eighties and meet his handsome dark-haired wife—not Maxwell but James Wright. Heavy glasses weighing above the close-cropped blond-gray bristles of a full beard, and everything he says is complimentary and mild, about others, about the book of mine he's read, complimentary but not fulsome, staring in restful silence at his clasped hands to remember another detail, then mentioning that, a tortured hint of ecstasy swimming into his eyes as they hit mine, so that I feel I'm here on a pilgrimage, its details unspoken. No drinks. Done with that, he says, though his wife has one. He asks to take a walk and we end in Gracie Square, in the same park where I walked once, then he chooses a bench and my eyes roll to locate things—This is the one I sat on, I believe, that day, alone.

"Tell me, Larry, what do you think—is it so?"

This is the reason for the night, I realize, and wonder if he's aware of Maxwell and my visit here or what it is he's asking, maybe a question about Michael, our mutual editor, or something about my wife or his, and I feel myself backing from this like a person gone flop on his butt trying to scramble uphill from the gravel slide starting to give way underfoot.

"So tell me, do you think Jesus is God?"

Back home, I feed the furnace, and just as I wonder how I'll keep up, I see John's rig down the road, a pickup-box trailer behind his pickup, both mounded with wood. He pulls around the drive to where I stand and gets out, his ponytail held tight down its length with multiple rubber bands, an anaconda emerging from under an orange watch cap, his huge weathered red face and frosted beard like Santa Claus. He grips my hand in a hard shake, mild blue eyes misting.

"Real bad about Joseph. I got calls from two people in your church and they both asked me to deliver you some wood so I brought you two loads. They said they'd pay."

. . .

I'm not sure I want to go upstairs, then do, legs weak, and see a maroon splotch over the hall carpet, a stain in his room, and, in the slant ceiling above the first flight of steps to the landing, the slug, imbedded. I pick at it and hear it roll inside in a gravelly rumble. The ceiling and walls of the stairwell are part of the original place, beaverboard, which I've meant to replace, so I dig with a knife at the point where I think the slug has rolled and it drops to the landing. Imbedded so loosely from running out of power after passing through his femur, then ricocheting at waist level off a blackboard on a side wall (*green*board, with a gouged streak in an angle pointing to its hit in the ceiling), with hairline grooves surrounding its tip where fibers of bone gripped it on its way through, a movie I must endure in slow motion, suffering through each second, as parents do.

Then I see the pistol hitting, its angle and ka-*blam*, and understand that with a slight change in inclination it could have struck—his spine, I think, in a hasty evasion meant to avoid what I'm afraid to think: *his head*. My legs go wobbly and I have to sit hard on the landing. I close my eyes and grip the slug, feeling it give off heat in my hand.

Down in the street from Farrar, Straus, & Giroux, the book done, my final changes added by hand, I start for the curb and feel reduced to baseless fabric, the substance bleeding from my legs, and then a gust of wind, a molecular gale that sets my hair erect but doesn't affect bystanders or people passing, rushes through me with a force I feel will fold me at the waist and fill me, a sail catapulting across the river to my wife and child, into the life I'm destined to lead. That life has just begun. I will remember every single day of it. *Mark that down*, I think. *I will do that and I will not relent*. It has all been gravy, as another writer said in a poem and, better, grace and gracious people put in my way, and yet more grace. I wring my hands as if washing them, wishing they were wings to lift me off in this wind streaming through me in a force I've never felt until now, or so I think, and then I think, *I'm launched*.

A Note About the Author

Larry Woiwode's fiction has appeared in *The Atlantic, Esquire, Harper's, The New Yorker,* and many other publications. His first novel, *What I'm Going to Do, I Think,* received the William Faulkner Foundation Award; his second, *Beyond the Bedroom Wall,* was a finalist for both the National Book Award and the National Book Critics' Circle Award. He has been a Guggenheim Fellow, a recipient of the Dos Passos Prize and, from the American Academy of Arts and Letters, the Medal of Merit. He is the poet laureate of North Dakota and lives there with his wife and children.

A Note on the Type

Rudolf Weiss designed this typeface in 1926 for the Bauer foundry of Frankfurt. Weiss is based on typefaces from the Italian Renaissance, and is one of the earliest contemporary serif types to have italics based on the chancery style of writing. The vertical strokes that are heavier at the top than at the bottom are unusual, and give Weiss a distinct beauty. Weiss is a legible text type and an elegant display face for headlines or titles. Weiss is a registered trademark of Bauer Types, S.A.